This Beekeeping log Book

Belongs to

_____

DEDICATION

This Beekeeping Log Book is dedicated to all the beekeepers out there who love working with bees.

You are my inspiration for producing books and I'm honored to be a part of keeping all of your important information and ideas organized all in one easy to find spot.

How to use this Beekeeping Log Book:

This ultimate beekeeping log notebook is a perfect way to track and record all your beekeeping activities. This unique Beekeeper's Journal is a great way to keep all of your important information all in one easy to find spot.

Each interior page includes prompts and space to record the following:

1. Colony Name - Write the name of the colony.

2. Date - Use the box provided to write in the date.

3. Weather Conditions - Stay on task using this space to record weather from sunny to rainy conditions.

4. Inspection - Rate the bee hive from 1-6 upon inspection.

5. Hive Number - Use the box to check mark the bee hive.

6. Productivity/Reproduction - Stay on task using the boxes for amount of honey, general population, amount of brood, and amount of space.

7. Behavior and Activities - Write the usual entering and ending activities, calm behavior when opening hive, bees bringing to hive, signs of robbery among the bees.

8. Health Status - Checklist if the bees seem weak or lazy, high amount of dead bees, queen bee is present, if she is identifiable, any infestation by ants, wax moth or any negative odor noted.

If you are new to beekeeping or have been at it for a while, this cute beekeeping journal is a must have! Can make a great useful gift for anyone that loves the hobby of beekeeping!

Have Fun!

COLONY NAME	
DATE	
TIME	

WEATHER CONDITIONS

🌡 _____ ☀ ⛅ 🌧 ⛈ ❄

🚩 _____ ☐ ☐ ☐ ☐ ☐

INSPECTION

HIVE NUMBER	①	②	③	④	⑤	⑥

PRODUCTIVITY & REPRODUCTION

AMOUNT OF HONEY						
GENERAL POPULATION						
AMOUNT OF BROOD						
AMOUNT OF SPACE						

BEHAVIOUR & ACTIVITIES

USUAL ENTERING AND EXITING ACTIVITY?						
CALM BEHAVIOUR WHEN OPENING HIVE?						
BEES BRINGING POLLEN INTO HIVE?						
SIGNS OF ROBBERY AMONG THE BEES?						

HEALTH STATUS

BEES SEEM WEAK OR LAZY?						
HIGH AMOUNT OF DEAD BEES?						
QUEEN BEE IS PRESENT / IDENTIFIABLE?						
INFESTATION BY ANTS / ANTS PRESENT?						
INFESTATION BY WAX MOTH / WAX MOTH PRESENT?						
NEGATIVE ODOUR NOTICEABLE?						

	COLONY NAME
	DATE
	TIME

WEATHER CONDITIONS

🌡 —— ☀ ⛅ 🌧 ⛈ ❄

🚩 —— ☐ ☐ ☐ ☐ ☐

INSPECTION

	HIVE NUMBER	①	②	③	④	⑤	⑥

PRODUCTIVITY & REPRODUCTION

	AMOUNT OF HONEY						
	GENERAL POPULATION						
	AMOUNT OF BROOD						
	AMOUNT OF SPACE						

BEHAVIOUR & ACTIVITIES

	USUAL ENTERING AND EXITING ACTIVITY?						
	CALM BEHAVIOUR WHEN OPENING HIVE?						
	BEES BRINGING POLLEN INTO HIVE?						
	SIGNS OF ROBBERY AMONG THE BEES?						

HEALTH STATUS

	BEES SEEM WEAK OR LAZY?						
	HIGH AMOUNT OF DEAD BEES?						
	QUEEN BEE IS PRESENT / IDENTIFIABLE?						
	INFESTATION BY ANTS / ANTS PRESENT?						
	INFESTATION BY WAX MOTH / WAX MOTH PRESENT?						
	NEGATIVE ODOUR NOTICEABLE?						

	COLONY NAME		WEATHER CONDITIONS

	COLONY NAME
	DATE
	TIME

WEATHER CONDITIONS

INSPECTION						
HIVE NUMBER	1	2	3	4	5	6

PRODUCTIVITY & REPRODUCTION						
AMOUNT OF HONEY						
GENERAL POPULATION						
AMOUNT OF BROOD						
AMOUNT OF SPACE						

BEHAVIOUR & ACTIVITIES						
USUAL ENTERING AND EXITING ACTIVITY?						
CALM BEHAVIOUR WHEN OPENING HIVE?						
BEES BRINGING POLLEN INTO HIVE?						
SIGNS OF ROBBERY AMONG THE BEES?						

HEALTH STATUS						
BEES SEEM WEAK OR LAZY?						
HIGH AMOUNT OF DEAD BEES?						
QUEEN BEE IS PRESENT / IDENTIFIABLE?						
INFESTATION BY ANTS / ANTS PRESENT?						
INFESTATION BY WAX MOTH / WAX MOTH PRESENT?						
NEGATIVE ODOUR NOTICEABLE?						

COLONY NAME	
DATE	
TIME	

WEATHER CONDITIONS

🌡	——	☀	⛅	☁	⛈	❄
🚩	——	☐	☐	☐	☐	☐

INSPECTION

HIVE NUMBER	①	②	③	④	⑤	⑥

PRODUCTIVITY & REPRODUCTION

AMOUNT OF HONEY						
GENERAL POPULATION						
AMOUNT OF BROOD						
AMOUNT OF SPACE						

BEHAVIOUR & ACTIVITIES

USUAL ENTERING AND EXITING ACTIVITY?						
CALM BEHAVIOUR WHEN OPENING HIVE?						
BEES BRINGING POLLEN INTO HIVE?						
SIGNS OF ROBBERY AMONG THE BEES?						

HEALTH STATUS

BEES SEEM WEAK OR LAZY?						
HIGH AMOUNT OF DEAD BEES?						
QUEEN BEE IS PRESENT / IDENTIFIABLE?						
INFESTATION BY ANTS / ANTS PRESENT?						
INFESTATION BY WAX MOTH / WAX MOTH PRESENT?						
NEGATIVE ODOUR NOTICEABLE?						

COLONY NAME	WEATHER CONDITIONS
DATE	
TIME	

INSPECTION						
HIVE NUMBER	①	②	③	④	⑤	⑥

PRODUCTIVITY & REPRODUCTION						
AMOUNT OF HONEY						
GENERAL POPULATION						
AMOUNT OF BROOD						
AMOUNT OF SPACE						

BEHAVIOUR & ACTIVITIES						
USUAL ENTERING AND EXITING ACTIVITY?						
CALM BEHAVIOUR WHEN OPENING HIVE?						
BEES BRINGING POLLEN INTO HIVE?						
SIGNS OF ROBBERY AMONG THE BEES?						

HEALTH STATUS						
BEES SEEM WEAK OR LAZY?						
HIGH AMOUNT OF DEAD BEES?						
QUEEN BEE IS PRESENT / IDENTIFIABLE?						
INFESTATION BY ANTS / ANTS PRESENT?						
INFESTATION BY WAX MOTH / WAX MOTH PRESENT?						
NEGATIVE ODOUR NOTICEABLE?						

COLONY NAME		WEATHER CONDITIONS	

COLONY NAME
DATE
TIME

WEATHER CONDITIONS

INSPECTION

HIVE NUMBER	1	2	3	4	5	6

PRODUCTIVITY & REPRODUCTION

AMOUNT OF HONEY						
GENERAL POPULATION						
AMOUNT OF BROOD						
AMOUNT OF SPACE						

BEHAVIOUR & ACTIVITIES

USUAL ENTERING AND EXITING ACTIVITY?						
CALM BEHAVIOUR WHEN OPENING HIVE?						
BEES BRINGING POLLEN INTO HIVE?						
SIGNS OF ROBBERY AMONG THE BEES?						

HEALTH STATUS

BEES SEEM WEAK OR LAZY?						
HIGH AMOUNT OF DEAD BEES?						
QUEEN BEE IS PRESENT / IDENTIFIABLE?						
INFESTATION BY ANTS / ANTS PRESENT?						
INFESTATION BY WAX MOTH / WAX MOTH PRESENT?						
NEGATIVE ODOUR NOTICEABLE?						

COLONY NAME		WEATHER CONDITIONS					
DATE		🌡 ___	☀	⛅	☁	⛈	❄
TIME		🚩 ___	☐	☐	☐	☐	☐

INSPECTION

HIVE NUMBER	①	②	③	④	⑤	⑥

PRODUCTIVITY & REPRODUCTION

AMOUNT OF HONEY						
GENERAL POPULATION						
AMOUNT OF BROOD						
AMOUNT OF SPACE						

BEHAVIOUR & ACTIVITIES

USUAL ENTERING AND EXITING ACTIVITY?						
CALM BEHAVIOUR WHEN OPENING HIVE?						
BEES BRINGING POLLEN INTO HIVE?						
SIGNS OF ROBBERY AMONG THE BEES?						

HEALTH STATUS

BEES SEEM WEAK OR LAZY?						
HIGH AMOUNT OF DEAD BEES?						
QUEEN BEE IS PRESENT / IDENTIFIABLE?						
INFESTATION BY ANTS / ANTS PRESENT?						
INFESTATION BY WAX MOTH / WAX MOTH PRESENT?						
NEGATIVE ODOUR NOTICEABLE?						

COLONY NAME		WEATHER CONDITIONS					
DATE		🌡 ___	☀	⛅	🌧	🌦	❄
TIME		🎏 ___	☐	☐	☐	☐	☐

INSPECTION

HIVE NUMBER	①	②	③	④	⑤	⑥

PRODUCTIVITY & REPRODUCTION

AMOUNT OF HONEY						
GENERAL POPULATION						
AMOUNT OF BROOD						
AMOUNT OF SPACE						

BEHAVIOUR & ACTIVITIES

USUAL ENTERING AND EXITING ACTIVITY?						
CALM BEHAVIOUR WHEN OPENING HIVE?						
BEES BRINGING POLLEN INTO HIVE?						
SIGNS OF ROBBERY AMONG THE BEES?						

HEALTH STATUS

BEES SEEM WEAK OR LAZY?						
HIGH AMOUNT OF DEAD BEES?						
QUEEN BEE IS PRESENT / IDENTIFIABLE?						
INFESTATION BY ANTS / ANTS PRESENT?						
INFESTATION BY WAX MOTH / WAX MOTH PRESENT?						
NEGATIVE ODOUR NOTICEABLE?						

COLONY NAME		WEATHER CONDITIONS					
DATE							
TIME							

INSPECTION

HIVE NUMBER	1	2	3	4	5	6

PRODUCTIVITY & REPRODUCTION

AMOUNT OF HONEY						
GENERAL POPULATION						
AMOUNT OF BROOD						
AMOUNT OF SPACE						

BEHAVIOUR & ACTIVITIES

USUAL ENTERING AND EXITING ACTIVITY?						
CALM BEHAVIOUR WHEN OPENING HIVE?						
BEES BRINGING POLLEN INTO HIVE?						
SIGNS OF ROBBERY AMONG THE BEES?						

HEALTH STATUS

BEES SEEM WEAK OR LAZY?						
HIGH AMOUNT OF DEAD BEES?						
QUEEN BEE IS PRESENT / IDENTIFIABLE?						
INFESTATION BY ANTS / ANTS PRESENT?						
INFESTATION BY WAX MOTH / WAX MOTH PRESENT?						
NEGATIVE ODOUR NOTICEABLE?						

	COLONY NAME	
	DATE	
	TIME	

WEATHER CONDITIONS

🌡	——	☀	⛅	🌧	⛈	❄
	——	☐	☐	☐	☐	☐

INSPECTION

HIVE NUMBER	①	②	③	④	⑤	⑥

PRODUCTIVITY & REPRODUCTION

AMOUNT OF HONEY						
GENERAL POPULATION						
AMOUNT OF BROOD						
AMOUNT OF SPACE						

BEHAVIOUR & ACTIVITIES

USUAL ENTERING AND EXITING ACTIVITY?						
CALM BEHAVIOUR WHEN OPENING HIVE?						
BEES BRINGING POLLEN INTO HIVE?						
SIGNS OF ROBBERY AMONG THE BEES?						

HEALTH STATUS

BEES SEEM WEAK OR LAZY?						
HIGH AMOUNT OF DEAD BEES?						
QUEEN BEE IS PRESENT / IDENTIFIABLE?						
INFESTATION BY ANTS / ANTS PRESENT?						
INFESTATION BY WAX MOTH / WAX MOTH PRESENT?						
NEGATIVE ODOUR NOTICEABLE?						

COLONY NAME	
DATE	
TIME	

WEATHER CONDITIONS

		☀	⛅	☁	⛈	❄
🌡	——					
🚩	——	☐	☐	☐	☐	☐

INSPECTION

HIVE NUMBER	①	②	③	④	⑤	⑥

PRODUCTIVITY & REPRODUCTION

AMOUNT OF HONEY						
GENERAL POPULATION						
AMOUNT OF BROOD						
AMOUNT OF SPACE						

BEHAVIOUR & ACTIVITIES

USUAL ENTERING AND EXITING ACTIVITY?						
CALM BEHAVIOUR WHEN OPENING HIVE?						
BEES BRINGING POLLEN INTO HIVE?						
SIGNS OF ROBBERY AMONG THE BEES?						

HEALTH STATUS

BEES SEEM WEAK OR LAZY?						
HIGH AMOUNT OF DEAD BEES?						
QUEEN BEE IS PRESENT / IDENTIFIABLE?						
INFESTATION BY ANTS / ANTS PRESENT?						
INFESTATION BY WAX MOTH / WAX MOTH PRESENT?						
NEGATIVE ODOUR NOTICEABLE?						

COLONY NAME		WEATHER CONDITIONS					
DATE		🌡 ___	☀	⛅	🌧	⛈	❄
TIME		🚩 ___	☐	☐	☐	☐	☐

INSPECTION						
🍯 HIVE NUMBER	①	②	③	④	⑤	⑥

PRODUCTIVITY & REPRODUCTION						
AMOUNT OF HONEY						
GENERAL POPULATION						
AMOUNT OF BROOD						
AMOUNT OF SPACE						

BEHAVIOUR & ACTIVITIES						
USUAL ENTERING AND EXITING ACTIVITY?						
CALM BEHAVIOUR WHEN OPENING HIVE?						
BEES BRINGING POLLEN INTO HIVE?						
SIGNS OF ROBBERY AMONG THE BEES?						

HEALTH STATUS						
BEES SEEM WEAK OR LAZY?						
HIGH AMOUNT OF DEAD BEES?						
QUEEN BEE IS PRESENT / IDENTIFIABLE?						
INFESTATION BY ANTS / ANTS PRESENT?						
INFESTATION BY WAX MOTH / WAX MOTH PRESENT?						
NEGATIVE ODOUR NOTICEABLE?						

COLONY NAME		WEATHER CONDITIONS					
DATE		🌡 ___	☀	⛅	☁	🌧	❄
TIME		🚩 ___	☐	☐	☐	☐	☐

INSPECTION

HIVE NUMBER	①	②	③	④	⑤	⑥

PRODUCTIVITY & REPRODUCTION

AMOUNT OF HONEY						
GENERAL POPULATION						
AMOUNT OF BROOD						
AMOUNT OF SPACE						

BEHAVIOUR & ACTIVITIES

USUAL ENTERING AND EXITING ACTIVITY?						
CALM BEHAVIOUR WHEN OPENING HIVE?						
BEES BRINGING POLLEN INTO HIVE?						
SIGNS OF ROBBERY AMONG THE BEES?						

HEALTH STATUS

BEES SEEM WEAK OR LAZY?						
HIGH AMOUNT OF DEAD BEES?						
QUEEN BEE IS PRESENT / IDENTIFIABLE?						
INFESTATION BY ANTS / ANTS PRESENT?						
INFESTATION BY WAX MOTH / WAX MOTH PRESENT?						
NEGATIVE ODOUR NOTICEABLE?						

COLONY NAME		WEATHER CONDITIONS
DATE		
TIME		

INSPECTION

HIVE NUMBER	①	②	③	④	⑤	⑥

PRODUCTIVITY & REPRODUCTION

AMOUNT OF HONEY						
GENERAL POPULATION						
AMOUNT OF BROOD						
AMOUNT OF SPACE						

BEHAVIOUR & ACTIVITIES

USUAL ENTERING AND EXITING ACTIVITY?						
CALM BEHAVIOUR WHEN OPENING HIVE?						
BEES BRINGING POLLEN INTO HIVE?						
SIGNS OF ROBBERY AMONG THE BEES?						

HEALTH STATUS

BEES SEEM WEAK OR LAZY?						
HIGH AMOUNT OF DEAD BEES?						
QUEEN BEE IS PRESENT / IDENTIFIABLE?						
INFESTATION BY ANTS / ANTS PRESENT?						
INFESTATION BY WAX MOTH / WAX MOTH PRESENT?						
NEGATIVE ODOUR NOTICEABLE?						

	COLONY NAME	
	DATE	
	TIME	

WEATHER CONDITIONS

	—	☀	⛅	🌧	⛈	❄
	—	☐	☐	☐	☐	☐

INSPECTION

HIVE NUMBER	①	②	③	④	⑤	⑥

PRODUCTIVITY & REPRODUCTION

AMOUNT OF HONEY						
GENERAL POPULATION						
AMOUNT OF BROOD						
AMOUNT OF SPACE						

BEHAVIOUR & ACTIVITIES

USUAL ENTERING AND EXITING ACTIVITY?						
CALM BEHAVIOUR WHEN OPENING HIVE?						
BEES BRINGING POLLEN INTO HIVE?						
SIGNS OF ROBBERY AMONG THE BEES?						

HEALTH STATUS

BEES SEEM WEAK OR LAZY?						
HIGH AMOUNT OF DEAD BEES?						
QUEEN BEE IS PRESENT / IDENTIFIABLE?						
INFESTATION BY ANTS / ANTS PRESENT?						
INFESTATION BY WAX MOTH / WAX MOTH PRESENT?						
NEGATIVE ODOUR NOTICEABLE?						

COLONY NAME		WEATHER CONDITIONS

COLONY NAME

DATE

TIME

WEATHER CONDITIONS

🌡 ——— ☀ ⛅ 🌧 ⛈ ❄

🚩 ——— ☐ ☐ ☐ ☐ ☐

INSPECTION

HIVE NUMBER	①	②	③	④	⑤	⑥

PRODUCTIVITY & REPRODUCTION

AMOUNT OF HONEY						
GENERAL POPULATION						
AMOUNT OF BROOD						
AMOUNT OF SPACE						

BEHAVIOUR & ACTIVITIES

USUAL ENTERING AND EXITING ACTIVITY?						
CALM BEHAVIOUR WHEN OPENING HIVE?						
BEES BRINGING POLLEN INTO HIVE?						
SIGNS OF ROBBERY AMONG THE BEES?						

HEALTH STATUS

BEES SEEM WEAK OR LAZY?						
HIGH AMOUNT OF DEAD BEES?						
QUEEN BEE IS PRESENT / IDENTIFIABLE?						
INFESTATION BY ANTS / ANTS PRESENT?						
INFESTATION BY WAX MOTH / WAX MOTH PRESENT?						
NEGATIVE ODOUR NOTICEABLE?						

COLONY NAME		WEATHER CONDITIONS					
DATE			☀	⛅	☁	🌧	❄
TIME			☐	☐	☐	☐	☐

INSPECTION

HIVE NUMBER	①	②	③	④	⑤	⑥

PRODUCTIVITY & REPRODUCTION

AMOUNT OF HONEY						
GENERAL POPULATION						
AMOUNT OF BROOD						
AMOUNT OF SPACE						

BEHAVIOUR & ACTIVITIES

USUAL ENTERING AND EXITING ACTIVITY?						
CALM BEHAVIOUR WHEN OPENING HIVE?						
BEES BRINGING POLLEN INTO HIVE?						
SIGNS OF ROBBERY AMONG THE BEES?						

HEALTH STATUS

BEES SEEM WEAK OR LAZY?						
HIGH AMOUNT OF DEAD BEES?						
QUEEN BEE IS PRESENT / IDENTIFIABLE?						
INFESTATION BY ANTS / ANTS PRESENT?						
INFESTATION BY WAX MOTH / WAX MOTH PRESENT?						
NEGATIVE ODOUR NOTICEABLE?						

COLONY NAME	
DATE	
TIME	

WEATHER CONDITIONS

🌡 ——   ☀ 🌤 🌧 ⛈ ❄

🎐 ——   ☐ ☐ ☐ ☐ ☐

INSPECTION

HIVE NUMBER	①	②	③	④	⑤	⑥

PRODUCTIVITY & REPRODUCTION

AMOUNT OF HONEY						
GENERAL POPULATION						
AMOUNT OF BROOD						
AMOUNT OF SPACE						

BEHAVIOUR & ACTIVITIES

USUAL ENTERING AND EXITING ACTIVITY?						
CALM BEHAVIOUR WHEN OPENING HIVE?						
BEES BRINGING POLLEN INTO HIVE?						
SIGNS OF ROBBERY AMONG THE BEES?						

HEALTH STATUS

BEES SEEM WEAK OR LAZY?						
HIGH AMOUNT OF DEAD BEES?						
QUEEN BEE IS PRESENT / IDENTIFIABLE?						
INFESTATION BY ANTS / ANTS PRESENT?						
INFESTATION BY WAX MOTH / WAX MOTH PRESENT?						
NEGATIVE ODOUR NOTICEABLE?						

COLONY NAME		WEATHER CONDITIONS

COLONY NAME
DATE
TIME

WEATHER CONDITIONS

INSPECTION						
HIVE NUMBER	①	②	③	④	⑤	⑥

PRODUCTIVITY & REPRODUCTION						
AMOUNT OF HONEY						
GENERAL POPULATION						
AMOUNT OF BROOD						
AMOUNT OF SPACE						

BEHAVIOUR & ACTIVITIES						
USUAL ENTERING AND EXITING ACTIVITY?						
CALM BEHAVIOUR WHEN OPENING HIVE?						
BEES BRINGING POLLEN INTO HIVE?						
SIGNS OF ROBBERY AMONG THE BEES?						

HEALTH STATUS						
BEES SEEM WEAK OR LAZY?						
HIGH AMOUNT OF DEAD BEES?						
QUEEN BEE IS PRESENT / IDENTIFIABLE?						
INFESTATION BY ANTS / ANTS PRESENT?						
INFESTATION BY WAX MOTH / WAX MOTH PRESENT?						
NEGATIVE ODOUR NOTICEABLE?						

COLONY NAME	
DATE	
TIME	

WEATHER CONDITIONS

🌡 —— ☀ 🌤 🌧 ⛈ ❄

🎐 —— ☐ ☐ ☐ ☐ ☐

INSPECTION

HIVE NUMBER	①	②	③	④	⑤	⑥

PRODUCTIVITY & REPRODUCTION

AMOUNT OF HONEY						
GENERAL POPULATION						
AMOUNT OF BROOD						
AMOUNT OF SPACE						

BEHAVIOUR & ACTIVITIES

USUAL ENTERING AND EXITING ACTIVITY?						
CALM BEHAVIOUR WHEN OPENING HIVE?						
BEES BRINGING POLLEN INTO HIVE?						
SIGNS OF ROBBERY AMONG THE BEES?						

HEALTH STATUS

BEES SEEM WEAK OR LAZY?						
HIGH AMOUNT OF DEAD BEES?						
QUEEN BEE IS PRESENT / IDENTIFIABLE?						
INFESTATION BY ANTS / ANTS PRESENT?						
INFESTATION BY WAX MOTH / WAX MOTH PRESENT?						
NEGATIVE ODOUR NOTICEABLE?						

COLONY NAME		WEATHER CONDITIONS
DATE		
TIME		

INSPECTION						
HIVE NUMBER	①	②	③	④	⑤	⑥

PRODUCTIVITY & REPRODUCTION						
AMOUNT OF HONEY						
GENERAL POPULATION						
AMOUNT OF BROOD						
AMOUNT OF SPACE						

BEHAVIOUR & ACTIVITIES						
USUAL ENTERING AND EXITING ACTIVITY?						
CALM BEHAVIOUR WHEN OPENING HIVE?						
BEES BRINGING POLLEN INTO HIVE?						
SIGNS OF ROBBERY AMONG THE BEES?						

HEALTH STATUS						
BEES SEEM WEAK OR LAZY?						
HIGH AMOUNT OF DEAD BEES?						
QUEEN BEE IS PRESENT / IDENTIFIABLE?						
INFESTATION BY ANTS / ANTS PRESENT?						
INFESTATION BY WAX MOTH / WAX MOTH PRESENT?						
NEGATIVE ODOUR NOTICEABLE?						

COLONY NAME		WEATHER CONDITIONS

COLONY NAME
DATE
TIME

INSPECTION

HIVE NUMBER	1	2	3	4	5	6

PRODUCTIVITY & REPRODUCTION

AMOUNT OF HONEY						
GENERAL POPULATION						
AMOUNT OF BROOD						
AMOUNT OF SPACE						

BEHAVIOUR & ACTIVITIES

USUAL ENTERING AND EXITING ACTIVITY?						
CALM BEHAVIOUR WHEN OPENING HIVE?						
BEES BRINGING POLLEN INTO HIVE?						
SIGNS OF ROBBERY AMONG THE BEES?						

HEALTH STATUS

BEES SEEM WEAK OR LAZY?						
HIGH AMOUNT OF DEAD BEES?						
QUEEN BEE IS PRESENT / IDENTIFIABLE?						
INFESTATION BY ANTS / ANTS PRESENT?						
INFESTATION BY WAX MOTH / WAX MOTH PRESENT?						
NEGATIVE ODOUR NOTICEABLE?						

COLONY NAME		WEATHER CONDITIONS					
DATE		🌡 ___	☀	⛅	🌧	⛈	❄
TIME		🚩 ___	☐	☐	☐	☐	☐

INSPECTION

HIVE NUMBER	1	2	3	4	5	6

PRODUCTIVITY & REPRODUCTION

AMOUNT OF HONEY						
GENERAL POPULATION						
AMOUNT OF BROOD						
AMOUNT OF SPACE						

BEHAVIOUR & ACTIVITIES

USUAL ENTERING AND EXITING ACTIVITY?						
CALM BEHAVIOUR WHEN OPENING HIVE?						
BEES BRINGING POLLEN INTO HIVE?						
SIGNS OF ROBBERY AMONG THE BEES?						

HEALTH STATUS

BEES SEEM WEAK OR LAZY?						
HIGH AMOUNT OF DEAD BEES?						
QUEEN BEE IS PRESENT / IDENTIFIABLE?						
INFESTATION BY ANTS / ANTS PRESENT?						
INFESTATION BY WAX MOTH / WAX MOTH PRESENT?						
NEGATIVE ODOUR NOTICEABLE?						

COLONY NAME		WEATHER CONDITIONS

COLONY NAME
DATE
TIME

WEATHER CONDITIONS

🌡 —— ☀ ⛅ ☁ 🌧 ❄
💨 —— ☐ ☐ ☐ ☐ ☐

INSPECTION

HIVE NUMBER	①	②	③	④	⑤	⑥

PRODUCTIVITY & REPRODUCTION

AMOUNT OF HONEY						
GENERAL POPULATION						
AMOUNT OF BROOD						
AMOUNT OF SPACE						

BEHAVIOUR & ACTIVITIES

USUAL ENTERING AND EXITING ACTIVITY?						
CALM BEHAVIOUR WHEN OPENING HIVE?						
BEES BRINGING POLLEN INTO HIVE?						
SIGNS OF ROBBERY AMONG THE BEES?						

HEALTH STATUS

BEES SEEM WEAK OR LAZY?						
HIGH AMOUNT OF DEAD BEES?						
QUEEN BEE IS PRESENT / IDENTIFIABLE?						
INFESTATION BY ANTS / ANTS PRESENT?						
INFESTATION BY WAX MOTH / WAX MOTH PRESENT?						
NEGATIVE ODOUR NOTICEABLE?						

COLONY NAME		WEATHER CONDITIONS
DATE		
TIME		

INSPECTION						
HIVE NUMBER	①	②	③	④	⑤	⑥

PRODUCTIVITY & REPRODUCTION						
AMOUNT OF HONEY						
GENERAL POPULATION						
AMOUNT OF BROOD						
AMOUNT OF SPACE						

BEHAVIOUR & ACTIVITIES						
USUAL ENTERING AND EXITING ACTIVITY?						
CALM BEHAVIOUR WHEN OPENING HIVE?						
BEES BRINGING POLLEN INTO HIVE?						
SIGNS OF ROBBERY AMONG THE BEES?						

HEALTH STATUS						
BEES SEEM WEAK OR LAZY?						
HIGH AMOUNT OF DEAD BEES?						
QUEEN BEE IS PRESENT / IDENTIFIABLE?						
INFESTATION BY ANTS / ANTS PRESENT?						
INFESTATION BY WAX MOTH / WAX MOTH PRESENT?						
NEGATIVE ODOUR NOTICEABLE?						

COLONY NAME	.
DATE	
TIME	

WEATHER CONDITIONS

🌡 ——   ☀ 🌤 🌧 ⛈ ❄

🚩 ——   ☐ ☐ ☐ ☐ ☐

INSPECTION

HIVE NUMBER	①	②	③	④	⑤	⑥

PRODUCTIVITY & REPRODUCTION

AMOUNT OF HONEY						
GENERAL POPULATION						
AMOUNT OF BROOD						
AMOUNT OF SPACE						

BEHAVIOUR & ACTIVITIES

USUAL ENTERING AND EXITING ACTIVITY?						
CALM BEHAVIOUR WHEN OPENING HIVE?						
BEES BRINGING POLLEN INTO HIVE?						
SIGNS OF ROBBERY AMONG THE BEES?						

HEALTH STATUS

BEES SEEM WEAK OR LAZY?						
HIGH AMOUNT OF DEAD BEES?						
QUEEN BEE IS PRESENT / IDENTIFIABLE?						
INFESTATION BY ANTS / ANTS PRESENT?						
INFESTATION BY WAX MOTH / WAX MOTH PRESENT?						
NEGATIVE ODOUR NOTICEABLE?						

		COLONY NAME				WEATHER CONDITIONS					

COLONY NAME / DATE / TIME

| COLONY NAME |
| DATE |
| TIME |

WEATHER CONDITIONS

Temperature ——   ☀ ⛅ 🌧 ⛈ ❄

Wind ——   ☐ ☐ ☐ ☐ ☐

INSPECTION

HIVE NUMBER	1	2	3	4	5	6

PRODUCTIVITY & REPRODUCTION

AMOUNT OF HONEY						
GENERAL POPULATION						
AMOUNT OF BROOD						
AMOUNT OF SPACE						

BEHAVIOUR & ACTIVITIES

USUAL ENTERING AND EXITING ACTIVITY?						
CALM BEHAVIOUR WHEN OPENING HIVE?						
BEES BRINGING POLLEN INTO HIVE?						
SIGNS OF ROBBERY AMONG THE BEES?						

HEALTH STATUS

BEES SEEM WEAK OR LAZY?						
HIGH AMOUNT OF DEAD BEES?						
QUEEN BEE IS PRESENT / IDENTIFIABLE?						
INFESTATION BY ANTS / ANTS PRESENT?						
INFESTATION BY WAX MOTH / WAX MOTH PRESENT?						
NEGATIVE ODOUR NOTICEABLE?						

COLONY NAME	WEATHER CONDITIONS

COLONY NAME

DATE

TIME

WEATHER CONDITIONS

🌡 —— ☀ ⛅ ☁ ⛈ ❄

🎐 —— ☐ ☐ ☐ ☐ ☐

INSPECTION

HIVE NUMBER	①	②	③	④	⑤	⑥

PRODUCTIVITY & REPRODUCTION

AMOUNT OF HONEY						
GENERAL POPULATION						
AMOUNT OF BROOD						
AMOUNT OF SPACE						

BEHAVIOUR & ACTIVITIES

USUAL ENTERING AND EXITING ACTIVITY?						
CALM BEHAVIOUR WHEN OPENING HIVE?						
BEES BRINGING POLLEN INTO HIVE?						
SIGNS OF ROBBERY AMONG THE BEES?						

HEALTH STATUS

BEES SEEM WEAK OR LAZY?						
HIGH AMOUNT OF DEAD BEES?						
QUEEN BEE IS PRESENT / IDENTIFIABLE?						
INFESTATION BY ANTS / ANTS PRESENT?						
INFESTATION BY WAX MOTH / WAX MOTH PRESENT?						
NEGATIVE ODOUR NOTICEABLE?						

COLONY NAME		WEATHER CONDITIONS					
DATE		🌡 ___	☀	⛅	🌧	⛈	❄
TIME		🚩 ___	☐	☐	☐	☐	☐

INSPECTION

HIVE NUMBER	①	②	③	④	⑤	⑥

PRODUCTIVITY & REPRODUCTION

AMOUNT OF HONEY						
GENERAL POPULATION						
AMOUNT OF BROOD						
AMOUNT OF SPACE						

BEHAVIOUR & ACTIVITIES

USUAL ENTERING AND EXITING ACTIVITY?						
CALM BEHAVIOUR WHEN OPENING HIVE?						
BEES BRINGING POLLEN INTO HIVE?						
SIGNS OF ROBBERY AMONG THE BEES?						

HEALTH STATUS

BEES SEEM WEAK OR LAZY?						
HIGH AMOUNT OF DEAD BEES?						
QUEEN BEE IS PRESENT / IDENTIFIABLE?						
INFESTATION BY ANTS / ANTS PRESENT?						
INFESTATION BY WAX MOTH / WAX MOTH PRESENT?						
NEGATIVE ODOUR NOTICEABLE?						

COLONY NAME		WEATHER CONDITIONS

COLONY NAME
DATE
TIME

WEATHER CONDITIONS

🌡 —— ☀ ⛅ 🌧 ⛈ ❄

🌬 —— ☐ ☐ ☐ ☐ ☐

INSPECTION						
HIVE NUMBER	①	②	③	④	⑤	⑥

PRODUCTIVITY & REPRODUCTION						
AMOUNT OF HONEY						
GENERAL POPULATION						
AMOUNT OF BROOD						
AMOUNT OF SPACE						

BEHAVIOUR & ACTIVITIES						
USUAL ENTERING AND EXITING ACTIVITY?						
CALM BEHAVIOUR WHEN OPENING HIVE?						
BEES BRINGING POLLEN INTO HIVE?						
SIGNS OF ROBBERY AMONG THE BEES?						

HEALTH STATUS						
BEES SEEM WEAK OR LAZY?						
HIGH AMOUNT OF DEAD BEES?						
QUEEN BEE IS PRESENT / IDENTIFIABLE?						
INFESTATION BY ANTS / ANTS PRESENT?						
INFESTATION BY WAX MOTH / WAX MOTH PRESENT?						
NEGATIVE ODOUR NOTICEABLE?						

COLONY NAME		WEATHER CONDITIONS					
DATE		🌡 ⎯⎯	☀	⛅	☁	🌧	❄
TIME		🚩 ⎯⎯	☐	☐	☐	☐	☐

INSPECTION

HIVE NUMBER	①	②	③	④	⑤	⑥

PRODUCTIVITY & REPRODUCTION

AMOUNT OF HONEY						
GENERAL POPULATION						
AMOUNT OF BROOD						
AMOUNT OF SPACE						

BEHAVIOUR & ACTIVITIES

USUAL ENTERING AND EXITING ACTIVITY?						
CALM BEHAVIOUR WHEN OPENING HIVE?						
BEES BRINGING POLLEN INTO HIVE?						
SIGNS OF ROBBERY AMONG THE BEES?						

HEALTH STATUS

BEES SEEM WEAK OR LAZY?						
HIGH AMOUNT OF DEAD BEES?						
QUEEN BEE IS PRESENT / IDENTIFIABLE?						
INFESTATION BY ANTS / ANTS PRESENT?						
INFESTATION BY WAX MOTH / WAX MOTH PRESENT?						
NEGATIVE ODOUR NOTICEABLE?						

	COLONY NAME	
	DATE	
	TIME	

WEATHER CONDITIONS

🌡 —— ☀ ⛅ ☁ ⛈ ❄

💨 —— ☐ ☐ ☐ ☐ ☐

INSPECTION

	HIVE NUMBER	①	②	③	④	⑤	⑥

PRODUCTIVITY & REPRODUCTION

	AMOUNT OF HONEY						
	GENERAL POPULATION						
	AMOUNT OF BROOD						
	AMOUNT OF SPACE						

BEHAVIOUR & ACTIVITIES

	USUAL ENTERING AND EXITING ACTIVITY?						
	CALM BEHAVIOUR WHEN OPENING HIVE?						
	BEES BRINGING POLLEN INTO HIVE?						
	SIGNS OF ROBBERY AMONG THE BEES?						

HEALTH STATUS

	BEES SEEM WEAK OR LAZY?						
	HIGH AMOUNT OF DEAD BEES?						
	QUEEN BEE IS PRESENT / IDENTIFIABLE?						
	INFESTATION BY ANTS / ANTS PRESENT?						
	INFESTATION BY WAX MOTH / WAX MOTH PRESENT?						
	NEGATIVE ODOUR NOTICEABLE?						

	COLONY NAME	
	DATE	
	TIME	

WEATHER CONDITIONS

🌡	—	☀	⛅	🌧	⛈	❄
🚩	—	☐	☐	☐	☐	☐

INSPECTION

	HIVE NUMBER	①	②	③	④	⑤	⑥

PRODUCTIVITY & REPRODUCTION

	AMOUNT OF HONEY						
	GENERAL POPULATION						
	AMOUNT OF BROOD						
	AMOUNT OF SPACE						

BEHAVIOUR & ACTIVITIES

	USUAL ENTERING AND EXITING ACTIVITY?						
	CALM BEHAVIOUR WHEN OPENING HIVE?						
	BEES BRINGING POLLEN INTO HIVE?						
	SIGNS OF ROBBERY AMONG THE BEES?						

HEALTH STATUS

	BEES SEEM WEAK OR LAZY?						
	HIGH AMOUNT OF DEAD BEES?						
	QUEEN BEE IS PRESENT / IDENTIFIABLE?						
	INFESTATION BY ANTS / ANTS PRESENT?						
	INFESTATION BY WAX MOTH / WAX MOTH PRESENT?						
	NEGATIVE ODOUR NOTICEABLE?						

	COLONY NAME	
📅	DATE	
🕐	TIME	

WEATHER CONDITIONS

🌡️ —— ☀️ ⛅ 🌧️ ⛈️ ❄️

🚩 —— ☐ ☐ ☐ ☐ ☐

INSPECTION						
🍯 HIVE NUMBER	①	②	③	④	⑤	⑥

PRODUCTIVITY & REPRODUCTION						
AMOUNT OF HONEY						
GENERAL POPULATION						
AMOUNT OF BROOD						
AMOUNT OF SPACE						

BEHAVIOUR & ACTIVITIES						
USUAL ENTERING AND EXITING ACTIVITY?						
CALM BEHAVIOUR WHEN OPENING HIVE?						
BEES BRINGING POLLEN INTO HIVE?						
SIGNS OF ROBBERY AMONG THE BEES?						

HEALTH STATUS						
BEES SEEM WEAK OR LAZY?						
HIGH AMOUNT OF DEAD BEES?						
QUEEN BEE IS PRESENT / IDENTIFIABLE?						
INFESTATION BY ANTS / ANTS PRESENT?						
INFESTATION BY WAX MOTH / WAX MOTH PRESENT?						
NEGATIVE ODOUR NOTICEABLE?						

COLONY NAME		WEATHER CONDITIONS					
DATE		🌡 ___	☀	⛅	☁	⛈	❄
TIME		🚩 ___	☐	☐	☐	☐	☐

INSPECTION

	HIVE NUMBER	1	2	3	4	5	6

PRODUCTIVITY & REPRODUCTION

	AMOUNT OF HONEY						
	GENERAL POPULATION						
	AMOUNT OF BROOD						
	AMOUNT OF SPACE						

BEHAVIOUR & ACTIVITIES

	USUAL ENTERING AND EXITING ACTIVITY?						
	CALM BEHAVIOUR WHEN OPENING HIVE?						
	BEES BRINGING POLLEN INTO HIVE?						
	SIGNS OF ROBBERY AMONG THE BEES?						

HEALTH STATUS

	BEES SEEM WEAK OR LAZY?						
	HIGH AMOUNT OF DEAD BEES?						
	QUEEN BEE IS PRESENT / IDENTIFIABLE?						
	INFESTATION BY ANTS / ANTS PRESENT?						
	INFESTATION BY WAX MOTH / WAX MOTH PRESENT?						
	NEGATIVE ODOUR NOTICEABLE?						

	COLONY NAME
	DATE
	TIME

WEATHER CONDITIONS

🌡 ___	☀	⛅	🌧	⛈	❄
🚩 ___	☐	☐	☐	☐	☐

INSPECTION

HIVE NUMBER	①	②	③	④	⑤	⑥

PRODUCTIVITY & REPRODUCTION

AMOUNT OF HONEY						
GENERAL POPULATION						
AMOUNT OF BROOD						
AMOUNT OF SPACE						

BEHAVIOUR & ACTIVITIES

USUAL ENTERING AND EXITING ACTIVITY?						
CALM BEHAVIOUR WHEN OPENING HIVE?						
BEES BRINGING POLLEN INTO HIVE?						
SIGNS OF ROBBERY AMONG THE BEES?						

HEALTH STATUS

BEES SEEM WEAK OR LAZY?						
HIGH AMOUNT OF DEAD BEES?						
QUEEN BEE IS PRESENT / IDENTIFIABLE?						
INFESTATION BY ANTS / ANTS PRESENT?						
INFESTATION BY WAX MOTH / WAX MOTH PRESENT?						
NEGATIVE ODOUR NOTICEABLE?						

COLONY NAME	
DATE	
TIME	

WEATHER CONDITIONS

🌡 ____ ☀ ⛅ 🌧 ⛈ ❄

🌬 ____ ☐ ☐ ☐ ☐ ☐

INSPECTION

HIVE NUMBER	①	②	③	④	⑤	⑥

PRODUCTIVITY & REPRODUCTION

AMOUNT OF HONEY						
GENERAL POPULATION						
AMOUNT OF BROOD						
AMOUNT OF SPACE						

BEHAVIOUR & ACTIVITIES

USUAL ENTERING AND EXITING ACTIVITY?						
CALM BEHAVIOUR WHEN OPENING HIVE?						
BEES BRINGING POLLEN INTO HIVE?						
SIGNS OF ROBBERY AMONG THE BEES?						

HEALTH STATUS

BEES SEEM WEAK OR LAZY?						
HIGH AMOUNT OF DEAD BEES?						
QUEEN BEE IS PRESENT / IDENTIFIABLE?						
INFESTATION BY ANTS / ANTS PRESENT?						
INFESTATION BY WAX MOTH / WAX MOTH PRESENT?						
NEGATIVE ODOUR NOTICEABLE?						

COLONY NAME		WEATHER CONDITIONS	

COLONY NAME

DATE

TIME

WEATHER CONDITIONS

🌡 —— ☀ ⛅ 🌧 ⛈ ❄

🎐 —— ☐ ☐ ☐ ☐ ☐

INSPECTION

HIVE NUMBER	①	②	③	④	⑤	⑥

PRODUCTIVITY & REPRODUCTION

AMOUNT OF HONEY						
GENERAL POPULATION						
AMOUNT OF BROOD						
AMOUNT OF SPACE						

BEHAVIOUR & ACTIVITIES

USUAL ENTERING AND EXITING ACTIVITY?						
CALM BEHAVIOUR WHEN OPENING HIVE?						
BEES BRINGING POLLEN INTO HIVE?						
SIGNS OF ROBBERY AMONG THE BEES?						

HEALTH STATUS

BEES SEEM WEAK OR LAZY?						
HIGH AMOUNT OF DEAD BEES?						
QUEEN BEE IS PRESENT / IDENTIFIABLE?						
INFESTATION BY ANTS / ANTS PRESENT?						
INFESTATION BY WAX MOTH / WAX MOTH PRESENT?						
NEGATIVE ODOUR NOTICEABLE?						

COLONY NAME		WEATHER CONDITIONS					
DATE		🌡 ___	☀	⛅	🌧	⛈	❄
TIME		🚩 ___	☐	☐	☐	☐	☐

INSPECTION

HIVE NUMBER	①	②	③	④	⑤	⑥

PRODUCTIVITY & REPRODUCTION

AMOUNT OF HONEY						
GENERAL POPULATION						
AMOUNT OF BROOD						
AMOUNT OF SPACE						

BEHAVIOUR & ACTIVITIES

USUAL ENTERING AND EXITING ACTIVITY?						
CALM BEHAVIOUR WHEN OPENING HIVE?						
BEES BRINGING POLLEN INTO HIVE?						
SIGNS OF ROBBERY AMONG THE BEES?						

HEALTH STATUS

BEES SEEM WEAK OR LAZY?						
HIGH AMOUNT OF DEAD BEES?						
QUEEN BEE IS PRESENT / IDENTIFIABLE?						
INFESTATION BY ANTS / ANTS PRESENT?						
INFESTATION BY WAX MOTH / WAX MOTH PRESENT?						
NEGATIVE ODOUR NOTICEABLE?						

	COLONY NAME		WEATHER CONDITIONS					
	DATE		🌡 ___	☀	⛅	☁	🌧	❄
	TIME		🚩 ___	☐	☐	☐	☐	☐

INSPECTION

	HIVE NUMBER	①	②	③	④	⑤	⑥

PRODUCTIVITY & REPRODUCTION

	AMOUNT OF HONEY						
	GENERAL POPULATION						
	AMOUNT OF BROOD						
	AMOUNT OF SPACE						

BEHAVIOUR & ACTIVITIES

	USUAL ENTERING AND EXITING ACTIVITY?						
	CALM BEHAVIOUR WHEN OPENING HIVE?						
	BEES BRINGING POLLEN INTO HIVE?						
	SIGNS OF ROBBERY AMONG THE BEES?						

HEALTH STATUS

	BEES SEEM WEAK OR LAZY?						
	HIGH AMOUNT OF DEAD BEES?						
	QUEEN BEE IS PRESENT / IDENTIFIABLE?						
	INFESTATION BY ANTS / ANTS PRESENT?						
	INFESTATION BY WAX MOTH / WAX MOTH PRESENT?						
	NEGATIVE ODOUR NOTICEABLE?						

COLONY NAME		WEATHER CONDITIONS					
DATE			☐	☐	☐	☐	☐
TIME							

INSPECTION

HIVE NUMBER	①	②	③	④	⑤	⑥

PRODUCTIVITY & REPRODUCTION

AMOUNT OF HONEY						
GENERAL POPULATION						
AMOUNT OF BROOD						
AMOUNT OF SPACE						

BEHAVIOUR & ACTIVITIES

USUAL ENTERING AND EXITING ACTIVITY?						
CALM BEHAVIOUR WHEN OPENING HIVE?						
BEES BRINGING POLLEN INTO HIVE?						
SIGNS OF ROBBERY AMONG THE BEES?						

HEALTH STATUS

BEES SEEM WEAK OR LAZY?						
HIGH AMOUNT OF DEAD BEES?						
QUEEN BEE IS PRESENT / IDENTIFIABLE?						
INFESTATION BY ANTS / ANTS PRESENT?						
INFESTATION BY WAX MOTH / WAX MOTH PRESENT?						
NEGATIVE ODOUR NOTICEABLE?						

COLONY NAME		WEATHER CONDITIONS					
DATE		🌡 ———	☀	⛅	🌧	⛈	❄
TIME		🚩 ———	☐	☐	☐	☐	☐

INSPECTION

HIVE NUMBER	①	②	③	④	⑤	⑥

PRODUCTIVITY & REPRODUCTION

AMOUNT OF HONEY						
GENERAL POPULATION						
AMOUNT OF BROOD						
AMOUNT OF SPACE						

BEHAVIOUR & ACTIVITIES

USUAL ENTERING AND EXITING ACTIVITY?						
CALM BEHAVIOUR WHEN OPENING HIVE?						
BEES BRINGING POLLEN INTO HIVE?						
SIGNS OF ROBBERY AMONG THE BEES?						

HEALTH STATUS

BEES SEEM WEAK OR LAZY?						
HIGH AMOUNT OF DEAD BEES?						
QUEEN BEE IS PRESENT / IDENTIFIABLE?						
INFESTATION BY ANTS / ANTS PRESENT?						
INFESTATION BY WAX MOTH / WAX MOTH PRESENT?						
NEGATIVE ODOUR NOTICEABLE?						

COLONY NAME	
DATE	
TIME	

WEATHER CONDITIONS

🌡	___	☀	⛅	🌧	⛈	❄
🎏	___	☐	☐	☐	☐	☐

INSPECTION

HIVE NUMBER	①	②	③	④	⑤	⑥

PRODUCTIVITY & REPRODUCTION

	1	2	3	4	5	6
AMOUNT OF HONEY						
GENERAL POPULATION						
AMOUNT OF BROOD						
AMOUNT OF SPACE						

BEHAVIOUR & ACTIVITIES

	1	2	3	4	5	6
USUAL ENTERING AND EXITING ACTIVITY?						
CALM BEHAVIOUR WHEN OPENING HIVE?						
BEES BRINGING POLLEN INTO HIVE?						
SIGNS OF ROBBERY AMONG THE BEES?						

HEALTH STATUS

	1	2	3	4	5	6
BEES SEEM WEAK OR LAZY?						
HIGH AMOUNT OF DEAD BEES?						
QUEEN BEE IS PRESENT / IDENTIFIABLE?						
INFESTATION BY ANTS / ANTS PRESENT?						
INFESTATION BY WAX MOTH / WAX MOTH PRESENT?						
NEGATIVE ODOUR NOTICEABLE?						

COLONY NAME		WEATHER CONDITIONS					
DATE		☀	⛅	🌧	⛈	❄	
TIME		☐	☐	☐	☐	☐	

INSPECTION

🐝 HIVE NUMBER	①	②	③	④	⑤	⑥

PRODUCTIVITY & REPRODUCTION

AMOUNT OF HONEY						
GENERAL POPULATION						
AMOUNT OF BROOD						
AMOUNT OF SPACE						

BEHAVIOUR & ACTIVITIES

USUAL ENTERING AND EXITING ACTIVITY?						
CALM BEHAVIOUR WHEN OPENING HIVE?						
BEES BRINGING POLLEN INTO HIVE?						
SIGNS OF ROBBERY AMONG THE BEES?						

HEALTH STATUS

BEES SEEM WEAK OR LAZY?						
HIGH AMOUNT OF DEAD BEES?						
QUEEN BEE IS PRESENT / IDENTIFIABLE?						
INFESTATION BY ANTS / ANTS PRESENT?						
INFESTATION BY WAX MOTH / WAX MOTH PRESENT?						
NEGATIVE ODOUR NOTICEABLE?						

	COLONY NAME	
	DATE	
	TIME	

WEATHER CONDITIONS

🌡	___	☀	⛅	☁	🌧	❄
🚩	___	☐	☐	☐	☐	☐

INSPECTION

	HIVE NUMBER	①	②	③	④	⑤	⑥

PRODUCTIVITY & REPRODUCTION

	AMOUNT OF HONEY						
	GENERAL POPULATION						
	AMOUNT OF BROOD						
	AMOUNT OF SPACE						

BEHAVIOUR & ACTIVITIES

	USUAL ENTERING AND EXITING ACTIVITY?						
	CALM BEHAVIOUR WHEN OPENING HIVE?						
	BEES BRINGING POLLEN INTO HIVE?						
	SIGNS OF ROBBERY AMONG THE BEES?						

HEALTH STATUS

	BEES SEEM WEAK OR LAZY?						
	HIGH AMOUNT OF DEAD BEES?						
	QUEEN BEE IS PRESENT / IDENTIFIABLE?						
	INFESTATION BY ANTS / ANTS PRESENT?						
	INFESTATION BY WAX MOTH / WAX MOTH PRESENT?						
	NEGATIVE ODOUR NOTICEABLE?						

COLONY NAME		WEATHER CONDITIONS					
DATE		🌡 ___	☀	⛅	☁	🌧	❄
TIME		🚩 ___	☐	☐	☐	☐	☐

INSPECTION							
HIVE NUMBER	①	②	③	④	⑤	⑥

PRODUCTIVITY & REPRODUCTION						
AMOUNT OF HONEY						
GENERAL POPULATION						
AMOUNT OF BROOD						
AMOUNT OF SPACE						

BEHAVIOUR & ACTIVITIES						
USUAL ENTERING AND EXITING ACTIVITY?						
CALM BEHAVIOUR WHEN OPENING HIVE?						
BEES BRINGING POLLEN INTO HIVE?						
SIGNS OF ROBBERY AMONG THE BEES?						

HEALTH STATUS						
BEES SEEM WEAK OR LAZY?						
HIGH AMOUNT OF DEAD BEES?						
QUEEN BEE IS PRESENT / IDENTIFIABLE?						
INFESTATION BY ANTS / ANTS PRESENT?						
INFESTATION BY WAX MOTH / WAX MOTH PRESENT?						
NEGATIVE ODOUR NOTICEABLE?						

COLONY NAME	
DATE	
TIME	

WEATHER CONDITIONS

🌡️ _____ ☀️ ⛅ 🌧️ ⛈️ ❄️

🚩 _____ ☐ ☐ ☐ ☐ ☐

INSPECTION

HIVE NUMBER	①	②	③	④	⑤	⑥

PRODUCTIVITY & REPRODUCTION

AMOUNT OF HONEY						
GENERAL POPULATION						
AMOUNT OF BROOD						
AMOUNT OF SPACE						

BEHAVIOUR & ACTIVITIES

USUAL ENTERING AND EXITING ACTIVITY?						
CALM BEHAVIOUR WHEN OPENING HIVE?						
BEES BRINGING POLLEN INTO HIVE?						
SIGNS OF ROBBERY AMONG THE BEES?						

HEALTH STATUS

BEES SEEM WEAK OR LAZY?						
HIGH AMOUNT OF DEAD BEES?						
QUEEN BEE IS PRESENT / IDENTIFIABLE?						
INFESTATION BY ANTS / ANTS PRESENT?						
INFESTATION BY WAX MOTH / WAX MOTH PRESENT?						
NEGATIVE ODOUR NOTICEABLE?						

COLONY NAME		WEATHER CONDITIONS

COLONY NAME

DATE

TIME

WEATHER CONDITIONS

INSPECTION						
HIVE NUMBER	①	②	③	④	⑤	⑥

PRODUCTIVITY & REPRODUCTION						
AMOUNT OF HONEY						
GENERAL POPULATION						
AMOUNT OF BROOD						
AMOUNT OF SPACE						

BEHAVIOUR & ACTIVITIES						
USUAL ENTERING AND EXITING ACTIVITY?						
CALM BEHAVIOUR WHEN OPENING HIVE?						
BEES BRINGING POLLEN INTO HIVE?						
SIGNS OF ROBBERY AMONG THE BEES?						

HEALTH STATUS						
BEES SEEM WEAK OR LAZY?						
HIGH AMOUNT OF DEAD BEES?						
QUEEN BEE IS PRESENT / IDENTIFIABLE?						
INFESTATION BY ANTS / ANTS PRESENT?						
INFESTATION BY WAX MOTH / WAX MOTH PRESENT?						
NEGATIVE ODOUR NOTICEABLE?						

COLONY NAME		WEATHER CONDITIONS					
DATE		🌡 ___	☀	⛅	🌧	⛈	❄
TIME		🚩 ___	☐	☐	☐	☐	☐

INSPECTION

HIVE NUMBER	1	2	3	4	5	6

PRODUCTIVITY & REPRODUCTION

AMOUNT OF HONEY						
GENERAL POPULATION						
AMOUNT OF BROOD						
AMOUNT OF SPACE						

BEHAVIOUR & ACTIVITIES

USUAL ENTERING AND EXITING ACTIVITY?						
CALM BEHAVIOUR WHEN OPENING HIVE?						
BEES BRINGING POLLEN INTO HIVE?						
SIGNS OF ROBBERY AMONG THE BEES?						

HEALTH STATUS

BEES SEEM WEAK OR LAZY?						
HIGH AMOUNT OF DEAD BEES?						
QUEEN BEE IS PRESENT / IDENTIFIABLE?						
INFESTATION BY ANTS / ANTS PRESENT?						
INFESTATION BY WAX MOTH / WAX MOTH PRESENT?						
NEGATIVE ODOUR NOTICEABLE?						

COLONY NAME		WEATHER CONDITIONS						
DATE		🌡 ___		☀	⛅	🌧	⛈	❄
TIME		🌬 ___		☐	☐	☐	☐	☐

INSPECTION

HIVE NUMBER	①	②	③	④	⑤	⑥

PRODUCTIVITY & REPRODUCTION

AMOUNT OF HONEY						
GENERAL POPULATION						
AMOUNT OF BROOD						
AMOUNT OF SPACE						

BEHAVIOUR & ACTIVITIES

USUAL ENTERING AND EXITING ACTIVITY?						
CALM BEHAVIOUR WHEN OPENING HIVE?						
BEES BRINGING POLLEN INTO HIVE?						
SIGNS OF ROBBERY AMONG THE BEES?						

HEALTH STATUS

BEES SEEM WEAK OR LAZY?						
HIGH AMOUNT OF DEAD BEES?						
QUEEN BEE IS PRESENT / IDENTIFIABLE?						
INFESTATION BY ANTS / ANTS PRESENT?						
INFESTATION BY WAX MOTH / WAX MOTH PRESENT?						
NEGATIVE ODOUR NOTICEABLE?						

COLONY NAME	
DATE	
TIME	

WEATHER CONDITIONS

🌡 _____ ☀ ⛅ 🌧 ⛈ ❄

🚩 _____ ☐ ☐ ☐ ☐ ☐

INSPECTION

HIVE NUMBER	①	②	③	④	⑤	⑥

PRODUCTIVITY & REPRODUCTION

AMOUNT OF HONEY						
GENERAL POPULATION						
AMOUNT OF BROOD						
AMOUNT OF SPACE						

BEHAVIOUR & ACTIVITIES

USUAL ENTERING AND EXITING ACTIVITY?						
CALM BEHAVIOUR WHEN OPENING HIVE?						
BEES BRINGING POLLEN INTO HIVE?						
SIGNS OF ROBBERY AMONG THE BEES?						

HEALTH STATUS

BEES SEEM WEAK OR LAZY?						
HIGH AMOUNT OF DEAD BEES?						
QUEEN BEE IS PRESENT / IDENTIFIABLE?						
INFESTATION BY ANTS / ANTS PRESENT?						
INFESTATION BY WAX MOTH / WAX MOTH PRESENT?						
NEGATIVE ODOUR NOTICEABLE?						

COLONY NAME	
DATE	
TIME	

WEATHER CONDITIONS

Temperature: —— ☀ ⛅ ☁ 🌧 ❄

Wind: —— ☐ ☐ ☐ ☐ ☐

INSPECTION

HIVE NUMBER	1	2	3	4	5	6

PRODUCTIVITY & REPRODUCTION

AMOUNT OF HONEY						
GENERAL POPULATION						
AMOUNT OF BROOD						
AMOUNT OF SPACE						

BEHAVIOUR & ACTIVITIES

USUAL ENTERING AND EXITING ACTIVITY?						
CALM BEHAVIOUR WHEN OPENING HIVE?						
BEES BRINGING POLLEN INTO HIVE?						
SIGNS OF ROBBERY AMONG THE BEES?						

HEALTH STATUS

BEES SEEM WEAK OR LAZY?						
HIGH AMOUNT OF DEAD BEES?						
QUEEN BEE IS PRESENT / IDENTIFIABLE?						
INFESTATION BY ANTS / ANTS PRESENT?						
INFESTATION BY WAX MOTH / WAX MOTH PRESENT?						
NEGATIVE ODOUR NOTICEABLE?						

COLONY NAME		WEATHER CONDITIONS					
DATE		🌡 ___	☀	⛅	🌧	⛈	❄
TIME		🚩 ___	☐	☐	☐	☐	☐

INSPECTION

HIVE NUMBER	①	②	③	④	⑤	⑥

PRODUCTIVITY & REPRODUCTION

AMOUNT OF HONEY						
GENERAL POPULATION						
AMOUNT OF BROOD						
AMOUNT OF SPACE						

BEHAVIOUR & ACTIVITIES

USUAL ENTERING AND EXITING ACTIVITY?						
CALM BEHAVIOUR WHEN OPENING HIVE?						
BEES BRINGING POLLEN INTO HIVE?						
SIGNS OF ROBBERY AMONG THE BEES?						

HEALTH STATUS

BEES SEEM WEAK OR LAZY?						
HIGH AMOUNT OF DEAD BEES?						
QUEEN BEE IS PRESENT / IDENTIFIABLE?						
INFESTATION BY ANTS / ANTS PRESENT?						
INFESTATION BY WAX MOTH / WAX MOTH PRESENT?						
NEGATIVE ODOUR NOTICEABLE?						

COLONY NAME		WEATHER CONDITIONS					
DATE		🌡 ——— ☀ ⛅ 🌧 ⛈ ❄					
TIME		🚩 ——— ☐ ☐ ☐ ☐ ☐					

INSPECTION

HIVE NUMBER	①	②	③	④	⑤	⑥

PRODUCTIVITY & REPRODUCTION

AMOUNT OF HONEY						
GENERAL POPULATION						
AMOUNT OF BROOD						
AMOUNT OF SPACE						

BEHAVIOUR & ACTIVITIES

USUAL ENTERING AND EXITING ACTIVITY?						
CALM BEHAVIOUR WHEN OPENING HIVE?						
BEES BRINGING POLLEN INTO HIVE?						
SIGNS OF ROBBERY AMONG THE BEES?						

HEALTH STATUS

BEES SEEM WEAK OR LAZY?						
HIGH AMOUNT OF DEAD BEES?						
QUEEN BEE IS PRESENT / IDENTIFIABLE?						
INFESTATION BY ANTS / ANTS PRESENT?						
INFESTATION BY WAX MOTH / WAX MOTH PRESENT?						
NEGATIVE ODOUR NOTICEABLE?						

COLONY NAME	WEATHER CONDITIONS

COLONY NAME
DATE
TIME

WEATHER CONDITIONS

INSPECTION						
HIVE NUMBER	①	②	③	④	⑤	⑥

PRODUCTIVITY & REPRODUCTION						
AMOUNT OF HONEY						
GENERAL POPULATION						
AMOUNT OF BROOD						
AMOUNT OF SPACE						

BEHAVIOUR & ACTIVITIES						
USUAL ENTERING AND EXITING ACTIVITY?						
CALM BEHAVIOUR WHEN OPENING HIVE?						
BEES BRINGING POLLEN INTO HIVE?						
SIGNS OF ROBBERY AMONG THE BEES?						

HEALTH STATUS						
BEES SEEM WEAK OR LAZY?						
HIGH AMOUNT OF DEAD BEES?						
QUEEN BEE IS PRESENT / IDENTIFIABLE?						
INFESTATION BY ANTS / ANTS PRESENT?						
INFESTATION BY WAX MOTH / WAX MOTH PRESENT?						
NEGATIVE ODOUR NOTICEABLE?						

COLONY NAME		WEATHER CONDITIONS					
DATE		🌡 ——	☀	⛅	🌧	⛈	❄
TIME		🚩 ——	☐	☐	☐	☐	☐

INSPECTION

		1	2	3	4	5	6
	HIVE NUMBER	①	②	③	④	⑤	⑥

PRODUCTIVITY & REPRODUCTION

	AMOUNT OF HONEY						
	GENERAL POPULATION						
	AMOUNT OF BROOD						
	AMOUNT OF SPACE						

BEHAVIOUR & ACTIVITIES

	USUAL ENTERING AND EXITING ACTIVITY?						
	CALM BEHAVIOUR WHEN OPENING HIVE?						
	BEES BRINGING POLLEN INTO HIVE?						
	SIGNS OF ROBBERY AMONG THE BEES?						

HEALTH STATUS

	BEES SEEM WEAK OR LAZY?						
	HIGH AMOUNT OF DEAD BEES?						
	QUEEN BEE IS PRESENT / IDENTIFIABLE?						
	INFESTATION BY ANTS / ANTS PRESENT?						
	INFESTATION BY WAX MOTH / WAX MOTH PRESENT?						
	NEGATIVE ODOUR NOTICEABLE?						

COLONY NAME		WEATHER CONDITIONS

	COLONY NAME
	DATE
	TIME

WEATHER CONDITIONS

INSPECTION						
HIVE NUMBER	①	②	③	④	⑤	⑥

PRODUCTIVITY & REPRODUCTION						
AMOUNT OF HONEY						
GENERAL POPULATION						
AMOUNT OF BROOD						
AMOUNT OF SPACE						

BEHAVIOUR & ACTIVITIES						
USUAL ENTERING AND EXITING ACTIVITY?						
CALM BEHAVIOUR WHEN OPENING HIVE?						
BEES BRINGING POLLEN INTO HIVE?						
SIGNS OF ROBBERY AMONG THE BEES?						

HEALTH STATUS						
BEES SEEM WEAK OR LAZY?						
HIGH AMOUNT OF DEAD BEES?						
QUEEN BEE IS PRESENT / IDENTIFIABLE?						
INFESTATION BY ANTS / ANTS PRESENT?						
INFESTATION BY WAX MOTH / WAX MOTH PRESENT?						
NEGATIVE ODOUR NOTICEABLE?						

COLONY NAME		WEATHER CONDITIONS	

COLONY NAME
DATE
TIME

WEATHER CONDITIONS

🌡 ____ ☀ ⛅ 🌧 ⛈ ❄

🚩 ____ ☐ ☐ ☐ ☐ ☐

INSPECTION

HIVE NUMBER	1	2	3	4	5	6

PRODUCTIVITY & REPRODUCTION

AMOUNT OF HONEY						
GENERAL POPULATION						
AMOUNT OF BROOD						
AMOUNT OF SPACE						

BEHAVIOUR & ACTIVITIES

USUAL ENTERING AND EXITING ACTIVITY?						
CALM BEHAVIOUR WHEN OPENING HIVE?						
BEES BRINGING POLLEN INTO HIVE?						
SIGNS OF ROBBERY AMONG THE BEES?						

HEALTH STATUS

BEES SEEM WEAK OR LAZY?						
HIGH AMOUNT OF DEAD BEES?						
QUEEN BEE IS PRESENT / IDENTIFIABLE?						
INFESTATION BY ANTS / ANTS PRESENT?						
INFESTATION BY WAX MOTH / WAX MOTH PRESENT?						
NEGATIVE ODOUR NOTICEABLE?						

COLONY NAME		WEATHER CONDITIONS				
DATE		🌡 ——— ☀ ⛅ 🌧 ⛈ ❄				
TIME		🚩 ——— ☐ ☐ ☐ ☐ ☐				

INSPECTION

HIVE NUMBER	①	②	③	④	⑤	⑥

PRODUCTIVITY & REPRODUCTION

AMOUNT OF HONEY						
GENERAL POPULATION						
AMOUNT OF BROOD						
AMOUNT OF SPACE						

BEHAVIOUR & ACTIVITIES

USUAL ENTERING AND EXITING ACTIVITY?						
CALM BEHAVIOUR WHEN OPENING HIVE?						
BEES BRINGING POLLEN INTO HIVE?						
SIGNS OF ROBBERY AMONG THE BEES?						

HEALTH STATUS

BEES SEEM WEAK OR LAZY?						
HIGH AMOUNT OF DEAD BEES?						
QUEEN BEE IS PRESENT / IDENTIFIABLE?						
INFESTATION BY ANTS / ANTS PRESENT?						
INFESTATION BY WAX MOTH / WAX MOTH PRESENT?						
NEGATIVE ODOUR NOTICEABLE?						

	COLONY NAME
	DATE
	TIME

WEATHER CONDITIONS

🌡 —— ☀ ⛅ 🌧 ⛈ ❄
🚩 —— ☐ ☐ ☐ ☐ ☐

INSPECTION

		1	2	3	4	5	6
	HIVE NUMBER	①	②	③	④	⑤	⑥

PRODUCTIVITY & REPRODUCTION

	AMOUNT OF HONEY						
	GENERAL POPULATION						
	AMOUNT OF BROOD						
	AMOUNT OF SPACE						

BEHAVIOUR & ACTIVITIES

	USUAL ENTERING AND EXITING ACTIVITY?						
	CALM BEHAVIOUR WHEN OPENING HIVE?						
	BEES BRINGING POLLEN INTO HIVE?						
	SIGNS OF ROBBERY AMONG THE BEES?						

HEALTH STATUS

	BEES SEEM WEAK OR LAZY?						
	HIGH AMOUNT OF DEAD BEES?						
	QUEEN BEE IS PRESENT / IDENTIFIABLE?						
	INFESTATION BY ANTS / ANTS PRESENT?						
	INFESTATION BY WAX MOTH / WAX MOTH PRESENT?						
	NEGATIVE ODOUR NOTICEABLE?						

	COLONY NAME		WEATHER CONDITIONS

	COLONY NAME
	DATE
	TIME

WEATHER CONDITIONS

🌡 _____ ☀ ⛅ 🌧 ⛈ ❄

🚩 _____ ☐ ☐ ☐ ☐ ☐

INSPECTION

	HIVE NUMBER	①	②	③	④	⑤	⑥

PRODUCTIVITY & REPRODUCTION

	AMOUNT OF HONEY						
	GENERAL POPULATION						
	AMOUNT OF BROOD						
	AMOUNT OF SPACE						

BEHAVIOUR & ACTIVITIES

	USUAL ENTERING AND EXITING ACTIVITY?						
	CALM BEHAVIOUR WHEN OPENING HIVE?						
	BEES BRINGING POLLEN INTO HIVE?						
	SIGNS OF ROBBERY AMONG THE BEES?						

HEALTH STATUS

	BEES SEEM WEAK OR LAZY?						
	HIGH AMOUNT OF DEAD BEES?						
	QUEEN BEE IS PRESENT / IDENTIFIABLE?						
	INFESTATION BY ANTS / ANTS PRESENT?						
	INFESTATION BY WAX MOTH / WAX MOTH PRESENT?						
	NEGATIVE ODOUR NOTICEABLE?						

COLONY NAME		WEATHER CONDITIONS					
DATE		🌡 ____	☀	⛅	☁	⛈	❄
TIME		🚩 ____	☐	☐	☐	☐	☐

INSPECTION

HIVE NUMBER	①	②	③	④	⑤	⑥

PRODUCTIVITY & REPRODUCTION

AMOUNT OF HONEY						
GENERAL POPULATION						
AMOUNT OF BROOD						
AMOUNT OF SPACE						

BEHAVIOUR & ACTIVITIES

USUAL ENTERING AND EXITING ACTIVITY?						
CALM BEHAVIOUR WHEN OPENING HIVE?						
BEES BRINGING POLLEN INTO HIVE?						
SIGNS OF ROBBERY AMONG THE BEES?						

HEALTH STATUS

BEES SEEM WEAK OR LAZY?						
HIGH AMOUNT OF DEAD BEES?						
QUEEN BEE IS PRESENT / IDENTIFIABLE?						
INFESTATION BY ANTS / ANTS PRESENT?						
INFESTATION BY WAX MOTH / WAX MOTH PRESENT?						
NEGATIVE ODOUR NOTICEABLE?						

COLONY NAME		WEATHER CONDITIONS					
DATE		🌡 —— ☀ ⛅ 🌧 ⛈ ❄					
TIME		🌬 —— ☐ ☐ ☐ ☐ ☐					

INSPECTION

HIVE NUMBER		①	②	③	④	⑤	⑥

PRODUCTIVITY & REPRODUCTION

AMOUNT OF HONEY						
GENERAL POPULATION						
AMOUNT OF BROOD						
AMOUNT OF SPACE						

BEHAVIOUR & ACTIVITIES

USUAL ENTERING AND EXITING ACTIVITY?						
CALM BEHAVIOUR WHEN OPENING HIVE?						
BEES BRINGING POLLEN INTO HIVE?						
SIGNS OF ROBBERY AMONG THE BEES?						

HEALTH STATUS

BEES SEEM WEAK OR LAZY?						
HIGH AMOUNT OF DEAD BEES?						
QUEEN BEE IS PRESENT / IDENTIFIABLE?						
INFESTATION BY ANTS / ANTS PRESENT?						
INFESTATION BY WAX MOTH / WAX MOTH PRESENT?						
NEGATIVE ODOUR NOTICEABLE?						

COLONY NAME	WEATHER CONDITIONS

COLONY NAME
DATE
TIME

WEATHER CONDITIONS

🌡 —— ☀ ⛅ 🌧 ⛈ ❄

🚩 —— ☐ ☐ ☐ ☐ ☐

INSPECTION

HIVE NUMBER	1	2	3	4	5	6

PRODUCTIVITY & REPRODUCTION

AMOUNT OF HONEY						
GENERAL POPULATION						
AMOUNT OF BROOD						
AMOUNT OF SPACE						

BEHAVIOUR & ACTIVITIES

USUAL ENTERING AND EXITING ACTIVITY?						
CALM BEHAVIOUR WHEN OPENING HIVE?						
BEES BRINGING POLLEN INTO HIVE?						
SIGNS OF ROBBERY AMONG THE BEES?						

HEALTH STATUS

BEES SEEM WEAK OR LAZY?						
HIGH AMOUNT OF DEAD BEES?						
QUEEN BEE IS PRESENT / IDENTIFIABLE?						
INFESTATION BY ANTS / ANTS PRESENT?						
INFESTATION BY WAX MOTH / WAX MOTH PRESENT?						
NEGATIVE ODOUR NOTICEABLE?						

COLONY NAME		WEATHER CONDITIONS					
DATE		🌡 ___	☀	⛅	🌧	⛈	❄
TIME		🧭 ___	☐	☐	☐	☐	☐

INSPECTION							
HIVE NUMBER		①	②	③	④	⑤	⑥

PRODUCTIVITY & REPRODUCTION						
AMOUNT OF HONEY						
GENERAL POPULATION						
AMOUNT OF BROOD						
AMOUNT OF SPACE						

BEHAVIOUR & ACTIVITIES						
USUAL ENTERING AND EXITING ACTIVITY?						
CALM BEHAVIOUR WHEN OPENING HIVE?						
BEES BRINGING POLLEN INTO HIVE?						
SIGNS OF ROBBERY AMONG THE BEES?						

HEALTH STATUS						
BEES SEEM WEAK OR LAZY?						
HIGH AMOUNT OF DEAD BEES?						
QUEEN BEE IS PRESENT / IDENTIFIABLE?						
INFESTATION BY ANTS / ANTS PRESENT?						
INFESTATION BY WAX MOTH / WAX MOTH PRESENT?						
NEGATIVE ODOUR NOTICEABLE?						

	COLONY NAME
	DATE
	TIME

WEATHER CONDITIONS

		☀	⛅	☁	🌧	❄
🌡	___					
🚩	___	☐	☐	☐	☐	☐

INSPECTION

		1	2	3	4	5	6
	HIVE NUMBER						

PRODUCTIVITY & REPRODUCTION

	AMOUNT OF HONEY						
	GENERAL POPULATION						
	AMOUNT OF BROOD						
	AMOUNT OF SPACE						

BEHAVIOUR & ACTIVITIES

	USUAL ENTERING AND EXITING ACTIVITY?						
	CALM BEHAVIOUR WHEN OPENING HIVE?						
	BEES BRINGING POLLEN INTO HIVE?						
	SIGNS OF ROBBERY AMONG THE BEES?						

HEALTH STATUS

	BEES SEEM WEAK OR LAZY?						
	HIGH AMOUNT OF DEAD BEES?						
	QUEEN BEE IS PRESENT / IDENTIFIABLE?						
	INFESTATION BY ANTS / ANTS PRESENT?						
	INFESTATION BY WAX MOTH / WAX MOTH PRESENT?						
	NEGATIVE ODOUR NOTICEABLE?						

COLONY NAME		WEATHER CONDITIONS					
DATE			☀	⛅	☁	⛈	❄
TIME			☐	☐	☐	☐	☐

INSPECTION

HIVE NUMBER	①	②	③	④	⑤	⑥

PRODUCTIVITY & REPRODUCTION

AMOUNT OF HONEY						
GENERAL POPULATION						
AMOUNT OF BROOD						
AMOUNT OF SPACE						

BEHAVIOUR & ACTIVITIES

USUAL ENTERING AND EXITING ACTIVITY?						
CALM BEHAVIOUR WHEN OPENING HIVE?						
BEES BRINGING POLLEN INTO HIVE?						
SIGNS OF ROBBERY AMONG THE BEES?						

HEALTH STATUS

BEES SEEM WEAK OR LAZY?						
HIGH AMOUNT OF DEAD BEES?						
QUEEN BEE IS PRESENT / IDENTIFIABLE?						
INFESTATION BY ANTS / ANTS PRESENT?						
INFESTATION BY WAX MOTH / WAX MOTH PRESENT?						
NEGATIVE ODOUR NOTICEABLE?						

COLONY NAME		WEATHER CONDITIONS					
DATE		🌡 ____	☀	⛅	☁	🌧	❄
TIME		🎏 ____	☐	☐	☐	☐	☐

INSPECTION

HIVE NUMBER	①	②	③	④	⑤	⑥

PRODUCTIVITY & REPRODUCTION

AMOUNT OF HONEY						
GENERAL POPULATION						
AMOUNT OF BROOD						
AMOUNT OF SPACE						

BEHAVIOUR & ACTIVITIES

USUAL ENTERING AND EXITING ACTIVITY?						
CALM BEHAVIOUR WHEN OPENING HIVE?						
BEES BRINGING POLLEN INTO HIVE?						
SIGNS OF ROBBERY AMONG THE BEES?						

HEALTH STATUS

BEES SEEM WEAK OR LAZY?						
HIGH AMOUNT OF DEAD BEES?						
QUEEN BEE IS PRESENT / IDENTIFIABLE?						
INFESTATION BY ANTS / ANTS PRESENT?						
INFESTATION BY WAX MOTH / WAX MOTH PRESENT?						
NEGATIVE ODOUR NOTICEABLE?						

COLONY NAME		WEATHER CONDITIONS					
DATE			☀	⛅	🌧	⛈	❄
TIME			☐	☐	☐	☐	☐

INSPECTION

HIVE NUMBER	①	②	③	④	⑤	⑥

PRODUCTIVITY & REPRODUCTION

AMOUNT OF HONEY						
GENERAL POPULATION						
AMOUNT OF BROOD						
AMOUNT OF SPACE						

BEHAVIOUR & ACTIVITIES

USUAL ENTERING AND EXITING ACTIVITY?						
CALM BEHAVIOUR WHEN OPENING HIVE?						
BEES BRINGING POLLEN INTO HIVE?						
SIGNS OF ROBBERY AMONG THE BEES?						

HEALTH STATUS

BEES SEEM WEAK OR LAZY?						
HIGH AMOUNT OF DEAD BEES?						
QUEEN BEE IS PRESENT / IDENTIFIABLE?						
INFESTATION BY ANTS / ANTS PRESENT?						
INFESTATION BY WAX MOTH / WAX MOTH PRESENT?						
NEGATIVE ODOUR NOTICEABLE?						

COLONY NAME		WEATHER CONDITIONS					
DATE		🌡 ___	☀	⛅	🌧	⛈	❄
TIME		🚩 ___	☐	☐	☐	☐	☐

INSPECTION

HIVE NUMBER	①	②	③	④	⑤	⑥

PRODUCTIVITY & REPRODUCTION

AMOUNT OF HONEY						
GENERAL POPULATION						
AMOUNT OF BROOD						
AMOUNT OF SPACE						

BEHAVIOUR & ACTIVITIES

USUAL ENTERING AND EXITING ACTIVITY?						
CALM BEHAVIOUR WHEN OPENING HIVE?						
BEES BRINGING POLLEN INTO HIVE?						
SIGNS OF ROBBERY AMONG THE BEES?						

HEALTH STATUS

BEES SEEM WEAK OR LAZY?						
HIGH AMOUNT OF DEAD BEES?						
QUEEN BEE IS PRESENT / IDENTIFIABLE?						
INFESTATION BY ANTS / ANTS PRESENT?						
INFESTATION BY WAX MOTH / WAX MOTH PRESENT?						
NEGATIVE ODOUR NOTICEABLE?						

COLONY NAME	
DATE	
TIME	

WEATHER CONDITIONS

🌡 _____ ☀ ⛅ ☁🌧 ⛈ ❄

🚩 _____ ☐ ☐ ☐ ☐ ☐

INSPECTION

HIVE NUMBER	①	②	③	④	⑤	⑥

PRODUCTIVITY & REPRODUCTION

AMOUNT OF HONEY						
GENERAL POPULATION						
AMOUNT OF BROOD						
AMOUNT OF SPACE						

BEHAVIOUR & ACTIVITIES

USUAL ENTERING AND EXITING ACTIVITY?						
CALM BEHAVIOUR WHEN OPENING HIVE?						
BEES BRINGING POLLEN INTO HIVE?						
SIGNS OF ROBBERY AMONG THE BEES?						

HEALTH STATUS

BEES SEEM WEAK OR LAZY?						
HIGH AMOUNT OF DEAD BEES?						
QUEEN BEE IS PRESENT / IDENTIFIABLE?						
INFESTATION BY ANTS / ANTS PRESENT?						
INFESTATION BY WAX MOTH / WAX MOTH PRESENT?						
NEGATIVE ODOUR NOTICEABLE?						

COLONY NAME		WEATHER CONDITIONS					
DATE		🌡 ___	☀	⛅	🌧	⛈	❄
TIME		🚩 ___	☐	☐	☐	☐	☐

INSPECTION

HIVE NUMBER	①	②	③	④	⑤	⑥

PRODUCTIVITY & REPRODUCTION

AMOUNT OF HONEY						
GENERAL POPULATION						
AMOUNT OF BROOD						
AMOUNT OF SPACE						

BEHAVIOUR & ACTIVITIES

USUAL ENTERING AND EXITING ACTIVITY?						
CALM BEHAVIOUR WHEN OPENING HIVE?						
BEES BRINGING POLLEN INTO HIVE?						
SIGNS OF ROBBERY AMONG THE BEES?						

HEALTH STATUS

BEES SEEM WEAK OR LAZY?						
HIGH AMOUNT OF DEAD BEES?						
QUEEN BEE IS PRESENT / IDENTIFIABLE?						
INFESTATION BY ANTS / ANTS PRESENT?						
INFESTATION BY WAX MOTH / WAX MOTH PRESENT?						
NEGATIVE ODOUR NOTICEABLE?						

	COLONY NAME		WEATHER CONDITIONS

COLONY NAME	
DATE	
TIME	

WEATHER CONDITIONS

🌡	___	☀	⛅	🌧	⛈	❄
🎏	___	☐	☐	☐	☐	☐

INSPECTION

HIVE NUMBER	①	②	③	④	⑤	⑥

PRODUCTIVITY & REPRODUCTION

AMOUNT OF HONEY						
GENERAL POPULATION						
AMOUNT OF BROOD						
AMOUNT OF SPACE						

BEHAVIOUR & ACTIVITIES

USUAL ENTERING AND EXITING ACTIVITY?						
CALM BEHAVIOUR WHEN OPENING HIVE?						
BEES BRINGING POLLEN INTO HIVE?						
SIGNS OF ROBBERY AMONG THE BEES?						

HEALTH STATUS

BEES SEEM WEAK OR LAZY?						
HIGH AMOUNT OF DEAD BEES?						
QUEEN BEE IS PRESENT / IDENTIFIABLE?						
INFESTATION BY ANTS / ANTS PRESENT?						
INFESTATION BY WAX MOTH / WAX MOTH PRESENT?						
NEGATIVE ODOUR NOTICEABLE?						

COLONY NAME		WEATHER CONDITIONS	
DATE		🌡 ___ ☀ ⛅ 🌧 ⛈ ❄	
TIME		🚩 ___ ☐ ☐ ☐ ☐ ☐	

INSPECTION

HIVE NUMBER	①	②	③	④	⑤	⑥

PRODUCTIVITY & REPRODUCTION

AMOUNT OF HONEY						
GENERAL POPULATION						
AMOUNT OF BROOD						
AMOUNT OF SPACE						

BEHAVIOUR & ACTIVITIES

USUAL ENTERING AND EXITING ACTIVITY?						
CALM BEHAVIOUR WHEN OPENING HIVE?						
BEES BRINGING POLLEN INTO HIVE?						
SIGNS OF ROBBERY AMONG THE BEES?						

HEALTH STATUS

BEES SEEM WEAK OR LAZY?						
HIGH AMOUNT OF DEAD BEES?						
QUEEN BEE IS PRESENT / IDENTIFIABLE?						
INFESTATION BY ANTS / ANTS PRESENT?						
INFESTATION BY WAX MOTH / WAX MOTH PRESENT?						
NEGATIVE ODOUR NOTICEABLE?						

COLONY NAME		WEATHER CONDITIONS					
DATE		☀	⛅	🌧	⛈	❄	
TIME		☐	☐	☐	☐	☐	

INSPECTION

	1	2	3	4	5	6
HIVE NUMBER	①	②	③	④	⑤	⑥

PRODUCTIVITY & REPRODUCTION

AMOUNT OF HONEY						
GENERAL POPULATION						
AMOUNT OF BROOD						
AMOUNT OF SPACE						

BEHAVIOUR & ACTIVITIES

USUAL ENTERING AND EXITING ACTIVITY?						
CALM BEHAVIOUR WHEN OPENING HIVE?						
BEES BRINGING POLLEN INTO HIVE?						
SIGNS OF ROBBERY AMONG THE BEES?						

HEALTH STATUS

BEES SEEM WEAK OR LAZY?						
HIGH AMOUNT OF DEAD BEES?						
QUEEN BEE IS PRESENT / IDENTIFIABLE?						
INFESTATION BY ANTS / ANTS PRESENT?						
INFESTATION BY WAX MOTH / WAX MOTH PRESENT?						
NEGATIVE ODOUR NOTICEABLE?						

	COLONY NAME
	DATE
	TIME

WEATHER CONDITIONS

🌡 _____ ☀ ⛅ 🌧 ⛈ ❄

🚩 _____ ☐ ☐ ☐ ☐ ☐

INSPECTION

	HIVE NUMBER	①	②	③	④	⑤	⑥

PRODUCTIVITY & REPRODUCTION

	AMOUNT OF HONEY						
	GENERAL POPULATION						
	AMOUNT OF BROOD						
	AMOUNT OF SPACE						

BEHAVIOUR & ACTIVITIES

	USUAL ENTERING AND EXITING ACTIVITY?						
	CALM BEHAVIOUR WHEN OPENING HIVE?						
	BEES BRINGING POLLEN INTO HIVE?						
	SIGNS OF ROBBERY AMONG THE BEES?						

HEALTH STATUS

	BEES SEEM WEAK OR LAZY?						
	HIGH AMOUNT OF DEAD BEES?						
	QUEEN BEE IS PRESENT / IDENTIFIABLE?						
	INFESTATION BY ANTS / ANTS PRESENT?						
	INFESTATION BY WAX MOTH / WAX MOTH PRESENT?						
	NEGATIVE ODOUR NOTICEABLE?						

	COLONY NAME
	DATE
	TIME

WEATHER CONDITIONS

	☀️	⛅	🌧️	⛈️	❄️
🌡️ ____					
____	☐	☐	☐	☐	☐

INSPECTION

HIVE NUMBER	①	②	③	④	⑤	⑥

PRODUCTIVITY & REPRODUCTION

AMOUNT OF HONEY						
GENERAL POPULATION						
AMOUNT OF BROOD						
AMOUNT OF SPACE						

BEHAVIOUR & ACTIVITIES

USUAL ENTERING AND EXITING ACTIVITY?						
CALM BEHAVIOUR WHEN OPENING HIVE?						
BEES BRINGING POLLEN INTO HIVE?						
SIGNS OF ROBBERY AMONG THE BEES?						

HEALTH STATUS

BEES SEEM WEAK OR LAZY?						
HIGH AMOUNT OF DEAD BEES?						
QUEEN BEE IS PRESENT / IDENTIFIABLE?						
INFESTATION BY ANTS / ANTS PRESENT?						
INFESTATION BY WAX MOTH / WAX MOTH PRESENT?						
NEGATIVE ODOUR NOTICEABLE?						

COLONY NAME	
DATE	
TIME	

WEATHER CONDITIONS

🌡 ____   ☀️ ⛅ 🌧 ⛈ ❄️

🚩 ____   ☐ ☐ ☐ ☐ ☐

INSPECTION

HIVE NUMBER	①	②	③	④	⑤	⑥

PRODUCTIVITY & REPRODUCTION

AMOUNT OF HONEY						
GENERAL POPULATION						
AMOUNT OF BROOD						
AMOUNT OF SPACE						

BEHAVIOUR & ACTIVITIES

USUAL ENTERING AND EXITING ACTIVITY?						
CALM BEHAVIOUR WHEN OPENING HIVE?						
BEES BRINGING POLLEN INTO HIVE?						
SIGNS OF ROBBERY AMONG THE BEES?						

HEALTH STATUS

BEES SEEM WEAK OR LAZY?						
HIGH AMOUNT OF DEAD BEES?						
QUEEN BEE IS PRESENT / IDENTIFIABLE?						
INFESTATION BY ANTS / ANTS PRESENT?						
INFESTATION BY WAX MOTH / WAX MOTH PRESENT?						
NEGATIVE ODOUR NOTICEABLE?						

COLONY NAME		WEATHER CONDITIONS					
DATE							
TIME							

INSPECTION

HIVE NUMBER	1	2	3	4	5	6

PRODUCTIVITY & REPRODUCTION

AMOUNT OF HONEY						
GENERAL POPULATION						
AMOUNT OF BROOD						
AMOUNT OF SPACE						

BEHAVIOUR & ACTIVITIES

USUAL ENTERING AND EXITING ACTIVITY?						
CALM BEHAVIOUR WHEN OPENING HIVE?						
BEES BRINGING POLLEN INTO HIVE?						
SIGNS OF ROBBERY AMONG THE BEES?						

HEALTH STATUS

BEES SEEM WEAK OR LAZY?						
HIGH AMOUNT OF DEAD BEES?						
QUEEN BEE IS PRESENT / IDENTIFIABLE?						
INFESTATION BY ANTS / ANTS PRESENT?						
INFESTATION BY WAX MOTH / WAX MOTH PRESENT?						
NEGATIVE ODOUR NOTICEABLE?						

	COLONY NAME	
	DATE	
	TIME	

WEATHER CONDITIONS

🌡	—	☀	⛅	🌧	⛈	❄
🚩	—	☐	☐	☐	☐	☐

INSPECTION

HIVE NUMBER	①	②	③	④	⑤	⑥

PRODUCTIVITY & REPRODUCTION

AMOUNT OF HONEY						
GENERAL POPULATION						
AMOUNT OF BROOD						
AMOUNT OF SPACE						

BEHAVIOUR & ACTIVITIES

USUAL ENTERING AND EXITING ACTIVITY?						
CALM BEHAVIOUR WHEN OPENING HIVE?						
BEES BRINGING POLLEN INTO HIVE?						
SIGNS OF ROBBERY AMONG THE BEES?						

HEALTH STATUS

BEES SEEM WEAK OR LAZY?						
HIGH AMOUNT OF DEAD BEES?						
QUEEN BEE IS PRESENT / IDENTIFIABLE?						
INFESTATION BY ANTS / ANTS PRESENT?						
INFESTATION BY WAX MOTH / WAX MOTH PRESENT?						
NEGATIVE ODOUR NOTICEABLE?						

	COLONY NAME			WEATHER CONDITIONS				

	COLONY NAME
	DATE
	TIME

WEATHER CONDITIONS

	INSPECTION						
HIVE NUMBER		1	2	3	4	5	6

PRODUCTIVITY & REPRODUCTION						
AMOUNT OF HONEY						
GENERAL POPULATION						
AMOUNT OF BROOD						
AMOUNT OF SPACE						

BEHAVIOUR & ACTIVITIES						
USUAL ENTERING AND EXITING ACTIVITY?						
CALM BEHAVIOUR WHEN OPENING HIVE?						
BEES BRINGING POLLEN INTO HIVE?						
SIGNS OF ROBBERY AMONG THE BEES?						

HEALTH STATUS						
BEES SEEM WEAK OR LAZY?						
HIGH AMOUNT OF DEAD BEES?						
QUEEN BEE IS PRESENT / IDENTIFIABLE?						
INFESTATION BY ANTS / ANTS PRESENT?						
INFESTATION BY WAX MOTH / WAX MOTH PRESENT?						
NEGATIVE ODOUR NOTICEABLE?						

COLONY NAME		WEATHER CONDITIONS
DATE		
TIME		

WEATHER CONDITIONS

☀ 🌤 🌧 ⛈ ❄
☐ ☐ ☐ ☐ ☐

INSPECTION

HIVE NUMBER	①	②	③	④	⑤	⑥

PRODUCTIVITY & REPRODUCTION

AMOUNT OF HONEY						
GENERAL POPULATION						
AMOUNT OF BROOD						
AMOUNT OF SPACE						

BEHAVIOUR & ACTIVITIES

USUAL ENTERING AND EXITING ACTIVITY?						
CALM BEHAVIOUR WHEN OPENING HIVE?						
BEES BRINGING POLLEN INTO HIVE?						
SIGNS OF ROBBERY AMONG THE BEES?						

HEALTH STATUS

BEES SEEM WEAK OR LAZY?						
HIGH AMOUNT OF DEAD BEES?						
QUEEN BEE IS PRESENT / IDENTIFIABLE?						
INFESTATION BY ANTS / ANTS PRESENT?						
INFESTATION BY WAX MOTH / WAX MOTH PRESENT?						
NEGATIVE ODOUR NOTICEABLE?						

COLONY NAME		WEATHER CONDITIONS					
DATE		☀	⛅	☁	🌧	❄	
TIME		☐	☐	☐	☐	☐	

INSPECTION							
HIVE NUMBER		①	②	③	④	⑤	⑥

PRODUCTIVITY & REPRODUCTION						
AMOUNT OF HONEY						
GENERAL POPULATION						
AMOUNT OF BROOD						
AMOUNT OF SPACE						

BEHAVIOUR & ACTIVITIES						
USUAL ENTERING AND EXITING ACTIVITY?						
CALM BEHAVIOUR WHEN OPENING HIVE?						
BEES BRINGING POLLEN INTO HIVE?						
SIGNS OF ROBBERY AMONG THE BEES?						

HEALTH STATUS						
BEES SEEM WEAK OR LAZY?						
HIGH AMOUNT OF DEAD BEES?						
QUEEN BEE IS PRESENT / IDENTIFIABLE?						
INFESTATION BY ANTS / ANTS PRESENT?						
INFESTATION BY WAX MOTH / WAX MOTH PRESENT?						
NEGATIVE ODOUR NOTICEABLE?						

COLONY NAME	WEATHER CONDITIONS
DATE	
TIME	

Weather: temperature ___ with icons ☀ ⛅ 🌧 ⛈ ❄
Wind: ___ with checkboxes ☐ ☐ ☐ ☐ ☐

INSPECTION

HIVE NUMBER	①	②	③	④	⑤	⑥

PRODUCTIVITY & REPRODUCTION

AMOUNT OF HONEY						
GENERAL POPULATION						
AMOUNT OF BROOD						
AMOUNT OF SPACE						

BEHAVIOUR & ACTIVITIES

USUAL ENTERING AND EXITING ACTIVITY?						
CALM BEHAVIOUR WHEN OPENING HIVE?						
BEES BRINGING POLLEN INTO HIVE?						
SIGNS OF ROBBERY AMONG THE BEES?						

HEALTH STATUS

BEES SEEM WEAK OR LAZY?						
HIGH AMOUNT OF DEAD BEES?						
QUEEN BEE IS PRESENT / IDENTIFIABLE?						
INFESTATION BY ANTS / ANTS PRESENT?						
INFESTATION BY WAX MOTH / WAX MOTH PRESENT?						
NEGATIVE ODOUR NOTICEABLE?						

COLONY NAME		WEATHER CONDITIONS	

INSPECTION

HIVE NUMBER	①	②	③	④	⑤	⑥

PRODUCTIVITY & REPRODUCTION

AMOUNT OF HONEY						
GENERAL POPULATION						
AMOUNT OF BROOD						
AMOUNT OF SPACE						

BEHAVIOUR & ACTIVITIES

USUAL ENTERING AND EXITING ACTIVITY?						
CALM BEHAVIOUR WHEN OPENING HIVE?						
BEES BRINGING POLLEN INTO HIVE?						
SIGNS OF ROBBERY AMONG THE BEES?						

HEALTH STATUS

BEES SEEM WEAK OR LAZY?						
HIGH AMOUNT OF DEAD BEES?						
QUEEN BEE IS PRESENT / IDENTIFIABLE?						
INFESTATION BY ANTS / ANTS PRESENT?						
INFESTATION BY WAX MOTH / WAX MOTH PRESENT?						
NEGATIVE ODOUR NOTICEABLE?						

DATE

TIME

COLONY NAME		WEATHER CONDITIONS					
DATE		🌡 ___	☀	⛅	🌧	⛈	❄
TIME		🚩 ___	☐	☐	☐	☐	☐

INSPECTION

HIVE NUMBER	①	②	③	④	⑤	⑥

PRODUCTIVITY & REPRODUCTION

AMOUNT OF HONEY						
GENERAL POPULATION						
AMOUNT OF BROOD						
AMOUNT OF SPACE						

BEHAVIOUR & ACTIVITIES

USUAL ENTERING AND EXITING ACTIVITY?						
CALM BEHAVIOUR WHEN OPENING HIVE?						
BEES BRINGING POLLEN INTO HIVE?						
SIGNS OF ROBBERY AMONG THE BEES?						

HEALTH STATUS

BEES SEEM WEAK OR LAZY?						
HIGH AMOUNT OF DEAD BEES?						
QUEEN BEE IS PRESENT / IDENTIFIABLE?						
INFESTATION BY ANTS / ANTS PRESENT?						
INFESTATION BY WAX MOTH / WAX MOTH PRESENT?						
NEGATIVE ODOUR NOTICEABLE?						

COLONY NAME		WEATHER CONDITIONS					
DATE		☀	⛅	🌧	⛈	❄	
TIME		☐	☐	☐	☐	☐	

INSPECTION

		①	②	③	④	⑤	⑥
	HIVE NUMBER	1	2	3	4	5	6

PRODUCTIVITY & REPRODUCTION

	AMOUNT OF HONEY						
	GENERAL POPULATION						
	AMOUNT OF BROOD						
	AMOUNT OF SPACE						

BEHAVIOUR & ACTIVITIES

	USUAL ENTERING AND EXITING ACTIVITY?						
	CALM BEHAVIOUR WHEN OPENING HIVE?						
	BEES BRINGING POLLEN INTO HIVE?						
	SIGNS OF ROBBERY AMONG THE BEES?						

HEALTH STATUS

	BEES SEEM WEAK OR LAZY?						
	HIGH AMOUNT OF DEAD BEES?						
	QUEEN BEE IS PRESENT / IDENTIFIABLE?						
	INFESTATION BY ANTS / ANTS PRESENT?						
	INFESTATION BY WAX MOTH / WAX MOTH PRESENT?						
	NEGATIVE ODOUR NOTICEABLE?						

	COLONY NAME
	DATE
	TIME

WEATHER CONDITIONS

	☀️	⛅	☁️	🌧️	❄️
🌡️ ——					
🚩 ——	☐	☐	☐	☐	☐

INSPECTION

HIVE NUMBER	①	②	③	④	⑤	⑥

PRODUCTIVITY & REPRODUCTION

AMOUNT OF HONEY						
GENERAL POPULATION						
AMOUNT OF BROOD						
AMOUNT OF SPACE						

BEHAVIOUR & ACTIVITIES

USUAL ENTERING AND EXITING ACTIVITY?						
CALM BEHAVIOUR WHEN OPENING HIVE?						
BEES BRINGING POLLEN INTO HIVE?						
SIGNS OF ROBBERY AMONG THE BEES?						

HEALTH STATUS

BEES SEEM WEAK OR LAZY?						
HIGH AMOUNT OF DEAD BEES?						
QUEEN BEE IS PRESENT / IDENTIFIABLE?						
INFESTATION BY ANTS / ANTS PRESENT?						
INFESTATION BY WAX MOTH / WAX MOTH PRESENT?						
NEGATIVE ODOUR NOTICEABLE?						

COLONY NAME		WEATHER CONDITIONS					
DATE		🌡 ___	☀	⛅	☁	🌧	❄
TIME		💨 ___	☐	☐	☐	☐	☐

INSPECTION

HIVE NUMBER	①	②	③	④	⑤	⑥

PRODUCTIVITY & REPRODUCTION

AMOUNT OF HONEY						
GENERAL POPULATION						
AMOUNT OF BROOD						
AMOUNT OF SPACE						

BEHAVIOUR & ACTIVITIES

USUAL ENTERING AND EXITING ACTIVITY?						
CALM BEHAVIOUR WHEN OPENING HIVE?						
BEES BRINGING POLLEN INTO HIVE?						
SIGNS OF ROBBERY AMONG THE BEES?						

HEALTH STATUS

BEES SEEM WEAK OR LAZY?						
HIGH AMOUNT OF DEAD BEES?						
QUEEN BEE IS PRESENT / IDENTIFIABLE?						
INFESTATION BY ANTS / ANTS PRESENT?						
INFESTATION BY WAX MOTH / WAX MOTH PRESENT?						
NEGATIVE ODOUR NOTICEABLE?						

COLONY NAME		WEATHER CONDITIONS					
DATE		🌡 ———	☀	⛅	🌧	⛈	❄
TIME		💨 ———	☐	☐	☐	☐	☐

INSPECTION

HIVE NUMBER	①	②	③	④	⑤	⑥

PRODUCTIVITY & REPRODUCTION

AMOUNT OF HONEY						
GENERAL POPULATION						
AMOUNT OF BROOD						
AMOUNT OF SPACE						

BEHAVIOUR & ACTIVITIES

USUAL ENTERING AND EXITING ACTIVITY?						
CALM BEHAVIOUR WHEN OPENING HIVE?						
BEES BRINGING POLLEN INTO HIVE?						
SIGNS OF ROBBERY AMONG THE BEES?						

HEALTH STATUS

BEES SEEM WEAK OR LAZY?						
HIGH AMOUNT OF DEAD BEES?						
QUEEN BEE IS PRESENT / IDENTIFIABLE?						
INFESTATION BY ANTS / ANTS PRESENT?						
INFESTATION BY WAX MOTH / WAX MOTH PRESENT?						
NEGATIVE ODOUR NOTICEABLE?						

COLONY NAME		WEATHER CONDITIONS					
DATE		🌡 ____	☀	⛅	☁	⛈	❄
TIME		🏴 ____	☐	☐	☐	☐	☐

INSPECTION

HIVE NUMBER	①	②	③	④	⑤	⑥

PRODUCTIVITY & REPRODUCTION

AMOUNT OF HONEY						
GENERAL POPULATION						
AMOUNT OF BROOD						
AMOUNT OF SPACE						

BEHAVIOUR & ACTIVITIES

USUAL ENTERING AND EXITING ACTIVITY?						
CALM BEHAVIOUR WHEN OPENING HIVE?						
BEES BRINGING POLLEN INTO HIVE?						
SIGNS OF ROBBERY AMONG THE BEES?						

HEALTH STATUS

BEES SEEM WEAK OR LAZY?						
HIGH AMOUNT OF DEAD BEES?						
QUEEN BEE IS PRESENT / IDENTIFIABLE?						
INFESTATION BY ANTS / ANTS PRESENT?						
INFESTATION BY WAX MOTH / WAX MOTH PRESENT?						
NEGATIVE ODOUR NOTICEABLE?						

COLONY NAME		WEATHER CONDITIONS					
DATE		🌡 ___	☀	⛅	🌧	⛈	❄
TIME		🚩 ___	☐	☐	☐	☐	☐

INSPECTION

HIVE NUMBER	①	②	③	④	⑤	⑥

PRODUCTIVITY & REPRODUCTION

AMOUNT OF HONEY						
GENERAL POPULATION						
AMOUNT OF BROOD						
AMOUNT OF SPACE						

BEHAVIOUR & ACTIVITIES

USUAL ENTERING AND EXITING ACTIVITY?						
CALM BEHAVIOUR WHEN OPENING HIVE?						
BEES BRINGING POLLEN INTO HIVE?						
SIGNS OF ROBBERY AMONG THE BEES?						

HEALTH STATUS

BEES SEEM WEAK OR LAZY?						
HIGH AMOUNT OF DEAD BEES?						
QUEEN BEE IS PRESENT / IDENTIFIABLE?						
INFESTATION BY ANTS / ANTS PRESENT?						
INFESTATION BY WAX MOTH / WAX MOTH PRESENT?						
NEGATIVE ODOUR NOTICEABLE?						

COLONY NAME		WEATHER CONDITIONS

COLONY NAME	
DATE	
TIME	

WEATHER CONDITIONS

🌡 ——

🎐 ——

☀ ☁ 🌧 ⛈ ❄

☐ ☐ ☐ ☐ ☐

INSPECTION

HIVE NUMBER	①	②	③	④	⑤	⑥

PRODUCTIVITY & REPRODUCTION

AMOUNT OF HONEY						
GENERAL POPULATION						
AMOUNT OF BROOD						
AMOUNT OF SPACE						

BEHAVIOUR & ACTIVITIES

USUAL ENTERING AND EXITING ACTIVITY?						
CALM BEHAVIOUR WHEN OPENING HIVE?						
BEES BRINGING POLLEN INTO HIVE?						
SIGNS OF ROBBERY AMONG THE BEES?						

HEALTH STATUS

BEES SEEM WEAK OR LAZY?						
HIGH AMOUNT OF DEAD BEES?						
QUEEN BEE IS PRESENT / IDENTIFIABLE?						
INFESTATION BY ANTS / ANTS PRESENT?						
INFESTATION BY WAX MOTH / WAX MOTH PRESENT?						
NEGATIVE ODOUR NOTICEABLE?						

	COLONY NAME	
	DATE	
	TIME	

WEATHER CONDITIONS

🌡 _____ ☀ ⛅ 🌧 ⛈ ❄

🎏 _____ ☐ ☐ ☐ ☐ ☐

INSPECTION

	HIVE NUMBER	1	2	3	4	5	6

PRODUCTIVITY & REPRODUCTION

	AMOUNT OF HONEY						
	GENERAL POPULATION						
	AMOUNT OF BROOD						
	AMOUNT OF SPACE						

BEHAVIOUR & ACTIVITIES

	USUAL ENTERING AND EXITING ACTIVITY?						
	CALM BEHAVIOUR WHEN OPENING HIVE?						
	BEES BRINGING POLLEN INTO HIVE?						
	SIGNS OF ROBBERY AMONG THE BEES?						

HEALTH STATUS

	BEES SEEM WEAK OR LAZY?						
	HIGH AMOUNT OF DEAD BEES?						
	QUEEN BEE IS PRESENT / IDENTIFIABLE?						
	INFESTATION BY ANTS / ANTS PRESENT?						
	INFESTATION BY WAX MOTH / WAX MOTH PRESENT?						
	NEGATIVE ODOUR NOTICEABLE?						

COLONY NAME		WEATHER CONDITIONS					
DATE		🌡 ———	☀	⛅	🌧	⛈	❄
TIME		💨 ———	☐	☐	☐	☐	☐

INSPECTION						
HIVE NUMBER	①	②	③	④	⑤	⑥

PRODUCTIVITY & REPRODUCTION						
AMOUNT OF HONEY						
GENERAL POPULATION						
AMOUNT OF BROOD						
AMOUNT OF SPACE						

BEHAVIOUR & ACTIVITIES						
USUAL ENTERING AND EXITING ACTIVITY?						
CALM BEHAVIOUR WHEN OPENING HIVE?						
BEES BRINGING POLLEN INTO HIVE?						
SIGNS OF ROBBERY AMONG THE BEES?						

HEALTH STATUS						
BEES SEEM WEAK OR LAZY?						
HIGH AMOUNT OF DEAD BEES?						
QUEEN BEE IS PRESENT / IDENTIFIABLE?						
INFESTATION BY ANTS / ANTS PRESENT?						
INFESTATION BY WAX MOTH / WAX MOTH PRESENT?						
NEGATIVE ODOUR NOTICEABLE?						

COLONY NAME		WEATHER CONDITIONS					
DATE		🌡 ___ ☀ ⛅ 🌧 ⛈ ❄					
TIME		🎐 ___ ☐ ☐ ☐ ☐ ☐					

INSPECTION							
🍯 HIVE NUMBER	①	②	③	④	⑤	⑥

PRODUCTIVITY & REPRODUCTION						
AMOUNT OF HONEY						
GENERAL POPULATION						
AMOUNT OF BROOD						
AMOUNT OF SPACE						

BEHAVIOUR & ACTIVITIES						
USUAL ENTERING AND EXITING ACTIVITY?						
CALM BEHAVIOUR WHEN OPENING HIVE?						
BEES BRINGING POLLEN INTO HIVE?						
SIGNS OF ROBBERY AMONG THE BEES?						

HEALTH STATUS						
BEES SEEM WEAK OR LAZY?						
HIGH AMOUNT OF DEAD BEES?						
QUEEN BEE IS PRESENT / IDENTIFIABLE?						
INFESTATION BY ANTS / ANTS PRESENT?						
INFESTATION BY WAX MOTH / WAX MOTH PRESENT?						
NEGATIVE ODOUR NOTICEABLE?						

COLONY NAME		WEATHER CONDITIONS					
DATE		🌡 ———	☀	⛅	☁	⛈	❄
TIME		🎏 ———	☐	☐	☐	☐	☐

INSPECTION							
HIVE NUMBER		①	②	③	④	⑤	⑥

PRODUCTIVITY & REPRODUCTION						
AMOUNT OF HONEY						
GENERAL POPULATION						
AMOUNT OF BROOD						
AMOUNT OF SPACE						

BEHAVIOUR & ACTIVITIES						
USUAL ENTERING AND EXITING ACTIVITY?						
CALM BEHAVIOUR WHEN OPENING HIVE?						
BEES BRINGING POLLEN INTO HIVE?						
SIGNS OF ROBBERY AMONG THE BEES?						

HEALTH STATUS						
BEES SEEM WEAK OR LAZY?						
HIGH AMOUNT OF DEAD BEES?						
QUEEN BEE IS PRESENT / IDENTIFIABLE?						
INFESTATION BY ANTS / ANTS PRESENT?						
INFESTATION BY WAX MOTH / WAX MOTH PRESENT?						
NEGATIVE ODOUR NOTICEABLE?						

COLONY NAME

DATE

TIME

WEATHER CONDITIONS

🌡 ___	☀	⛅	🌧	⛈	❄
🚩 ___	☐	☐	☐	☐	☐

INSPECTION

HIVE NUMBER	①	②	③	④	⑤	⑥

PRODUCTIVITY & REPRODUCTION

AMOUNT OF HONEY						
GENERAL POPULATION						
AMOUNT OF BROOD						
AMOUNT OF SPACE						

BEHAVIOUR & ACTIVITIES

USUAL ENTERING AND EXITING ACTIVITY?						
CALM BEHAVIOUR WHEN OPENING HIVE?						
BEES BRINGING POLLEN INTO HIVE?						
SIGNS OF ROBBERY AMONG THE BEES?						

HEALTH STATUS

BEES SEEM WEAK OR LAZY?						
HIGH AMOUNT OF DEAD BEES?						
QUEEN BEE IS PRESENT / IDENTIFIABLE?						
INFESTATION BY ANTS / ANTS PRESENT?						
INFESTATION BY WAX MOTH / WAX MOTH PRESENT?						
NEGATIVE ODOUR NOTICEABLE?						

COLONY NAME	WEATHER CONDITIONS					
DATE	🌡 ____	☀	⛅	☁	🌧	❄
TIME	💨 ____	☐	☐	☐	☐	☐

INSPECTION

HIVE NUMBER	1	2	3	4	5	6

PRODUCTIVITY & REPRODUCTION

AMOUNT OF HONEY						
GENERAL POPULATION						
AMOUNT OF BROOD						
AMOUNT OF SPACE						

BEHAVIOUR & ACTIVITIES

USUAL ENTERING AND EXITING ACTIVITY?						
CALM BEHAVIOUR WHEN OPENING HIVE?						
BEES BRINGING POLLEN INTO HIVE?						
SIGNS OF ROBBERY AMONG THE BEES?						

HEALTH STATUS

BEES SEEM WEAK OR LAZY?						
HIGH AMOUNT OF DEAD BEES?						
QUEEN BEE IS PRESENT / IDENTIFIABLE?						
INFESTATION BY ANTS / ANTS PRESENT?						
INFESTATION BY WAX MOTH / WAX MOTH PRESENT?						
NEGATIVE ODOUR NOTICEABLE?						

COLONY NAME		WEATHER CONDITIONS				
DATE		🌡 ____ ☀ ⛅ 🌧 ⛈ ❄				
TIME		🎏 ____ ☐ ☐ ☐ ☐ ☐				

INSPECTION						
HIVE NUMBER	①	②	③	④	⑤	⑥

PRODUCTIVITY & REPRODUCTION						
AMOUNT OF HONEY						
GENERAL POPULATION						
AMOUNT OF BROOD						
AMOUNT OF SPACE						

BEHAVIOUR & ACTIVITIES						
USUAL ENTERING AND EXITING ACTIVITY?						
CALM BEHAVIOUR WHEN OPENING HIVE?						
BEES BRINGING POLLEN INTO HIVE?						
SIGNS OF ROBBERY AMONG THE BEES?						

HEALTH STATUS						
BEES SEEM WEAK OR LAZY?						
HIGH AMOUNT OF DEAD BEES?						
QUEEN BEE IS PRESENT / IDENTIFIABLE?						
INFESTATION BY ANTS / ANTS PRESENT?						
INFESTATION BY WAX MOTH / WAX MOTH PRESENT?						
NEGATIVE ODOUR NOTICEABLE?						

COLONY NAME	
DATE	
TIME	

WEATHER CONDITIONS

🌡 ———	☀	⛅	☁	🌧	❄
🎏 ———	☐	☐	☐	☐	☐

INSPECTION

HIVE NUMBER	①	②	③	④	⑤	⑥

PRODUCTIVITY & REPRODUCTION

AMOUNT OF HONEY						
GENERAL POPULATION						
AMOUNT OF BROOD						
AMOUNT OF SPACE						

BEHAVIOUR & ACTIVITIES

USUAL ENTERING AND EXITING ACTIVITY?						
CALM BEHAVIOUR WHEN OPENING HIVE?						
BEES BRINGING POLLEN INTO HIVE?						
SIGNS OF ROBBERY AMONG THE BEES?						

HEALTH STATUS

BEES SEEM WEAK OR LAZY?						
HIGH AMOUNT OF DEAD BEES?						
QUEEN BEE IS PRESENT / IDENTIFIABLE?						
INFESTATION BY ANTS / ANTS PRESENT?						
INFESTATION BY WAX MOTH / WAX MOTH PRESENT?						
NEGATIVE ODOUR NOTICEABLE?						

	COLONY NAME
	DATE
	TIME

WEATHER CONDITIONS

		☀	⛅	🌧	⛈	❄
🌡	____					
🎐	____	☐	☐	☐	☐	☐

INSPECTION

	HIVE NUMBER	①	②	③	④	⑤	⑥

PRODUCTIVITY & REPRODUCTION

	AMOUNT OF HONEY						
	GENERAL POPULATION						
	AMOUNT OF BROOD						
	AMOUNT OF SPACE						

BEHAVIOUR & ACTIVITIES

	USUAL ENTERING AND EXITING ACTIVITY?						
	CALM BEHAVIOUR WHEN OPENING HIVE?						
	BEES BRINGING POLLEN INTO HIVE?						
	SIGNS OF ROBBERY AMONG THE BEES?						

HEALTH STATUS

	BEES SEEM WEAK OR LAZY?						
	HIGH AMOUNT OF DEAD BEES?						
	QUEEN BEE IS PRESENT / IDENTIFIABLE?						
	INFESTATION BY ANTS / ANTS PRESENT?						
	INFESTATION BY WAX MOTH / WAX MOTH PRESENT?						
	NEGATIVE ODOUR NOTICEABLE?						

COLONY NAME		WEATHER CONDITIONS					
DATE		🌡 ———	☀	⛅	☁	🌧	❄
TIME		🚩 ———	☐	☐	☐	☐	☐

INSPECTION

HIVE NUMBER		①	②	③	④	⑤	⑥

PRODUCTIVITY & REPRODUCTION

AMOUNT OF HONEY						
GENERAL POPULATION						
AMOUNT OF BROOD						
AMOUNT OF SPACE						

BEHAVIOUR & ACTIVITIES

USUAL ENTERING AND EXITING ACTIVITY?						
CALM BEHAVIOUR WHEN OPENING HIVE?						
BEES BRINGING POLLEN INTO HIVE?						
SIGNS OF ROBBERY AMONG THE BEES?						

HEALTH STATUS

BEES SEEM WEAK OR LAZY?						
HIGH AMOUNT OF DEAD BEES?						
QUEEN BEE IS PRESENT / IDENTIFIABLE?						
INFESTATION BY ANTS / ANTS PRESENT?						
INFESTATION BY WAX MOTH / WAX MOTH PRESENT?						
NEGATIVE ODOUR NOTICEABLE?						

COLONY NAME		WEATHER CONDITIONS					
DATE							
TIME							

INSPECTION

HIVE NUMBER	1	2	3	4	5	6

PRODUCTIVITY & REPRODUCTION

AMOUNT OF HONEY						
GENERAL POPULATION						
AMOUNT OF BROOD						
AMOUNT OF SPACE						

BEHAVIOUR & ACTIVITIES

USUAL ENTERING AND EXITING ACTIVITY?						
CALM BEHAVIOUR WHEN OPENING HIVE?						
BEES BRINGING POLLEN INTO HIVE?						
SIGNS OF ROBBERY AMONG THE BEES?						

HEALTH STATUS

BEES SEEM WEAK OR LAZY?						
HIGH AMOUNT OF DEAD BEES?						
QUEEN BEE IS PRESENT / IDENTIFIABLE?						
INFESTATION BY ANTS / ANTS PRESENT?						
INFESTATION BY WAX MOTH / WAX MOTH PRESENT?						
NEGATIVE ODOUR NOTICEABLE?						

COLONY NAME		WEATHER CONDITIONS						
DATE		🌡 ___		☀	⛅	🌧	⛈	❄
TIME		🪁 ___		☐	☐	☐	☐	☐

INSPECTION

		①	②	③	④	⑤	⑥
HIVE NUMBER		1	2	3	4	5	6

PRODUCTIVITY & REPRODUCTION

AMOUNT OF HONEY						
GENERAL POPULATION						
AMOUNT OF BROOD						
AMOUNT OF SPACE						

BEHAVIOUR & ACTIVITIES

USUAL ENTERING AND EXITING ACTIVITY?						
CALM BEHAVIOUR WHEN OPENING HIVE?						
BEES BRINGING POLLEN INTO HIVE?						
SIGNS OF ROBBERY AMONG THE BEES?						

HEALTH STATUS

BEES SEEM WEAK OR LAZY?						
HIGH AMOUNT OF DEAD BEES?						
QUEEN BEE IS PRESENT / IDENTIFIABLE?						
INFESTATION BY ANTS / ANTS PRESENT?						
INFESTATION BY WAX MOTH / WAX MOTH PRESENT?						
NEGATIVE ODOUR NOTICEABLE?						

COLONY NAME	WEATHER CONDITIONS					
DATE	🌡 ___	☀	🌤	☁	🌧	❄
TIME	🚩 ___	☐	☐	☐	☐	☐

INSPECTION						
HIVE NUMBER	①	②	③	④	⑤	⑥

PRODUCTIVITY & REPRODUCTION						
AMOUNT OF HONEY						
GENERAL POPULATION						
AMOUNT OF BROOD						
AMOUNT OF SPACE						

BEHAVIOUR & ACTIVITIES						
USUAL ENTERING AND EXITING ACTIVITY?						
CALM BEHAVIOUR WHEN OPENING HIVE?						
BEES BRINGING POLLEN INTO HIVE?						
SIGNS OF ROBBERY AMONG THE BEES?						

HEALTH STATUS						
BEES SEEM WEAK OR LAZY?						
HIGH AMOUNT OF DEAD BEES?						
QUEEN BEE IS PRESENT / IDENTIFIABLE?						
INFESTATION BY ANTS / ANTS PRESENT?						
INFESTATION BY WAX MOTH / WAX MOTH PRESENT?						
NEGATIVE ODOUR NOTICEABLE?						

COLONY NAME	WEATHER CONDITIONS					
DATE	🌡 ___	☀	⛅	☁	🌧	❄
TIME	🎏 ___	☐	☐	☐	☐	☐

INSPECTION

	1	2	3	4	5	6
HIVE NUMBER	①	②	③	④	⑤	⑥

PRODUCTIVITY & REPRODUCTION

AMOUNT OF HONEY						
GENERAL POPULATION						
AMOUNT OF BROOD						
AMOUNT OF SPACE						

BEHAVIOUR & ACTIVITIES

USUAL ENTERING AND EXITING ACTIVITY?						
CALM BEHAVIOUR WHEN OPENING HIVE?						
BEES BRINGING POLLEN INTO HIVE?						
SIGNS OF ROBBERY AMONG THE BEES?						

HEALTH STATUS

BEES SEEM WEAK OR LAZY?						
HIGH AMOUNT OF DEAD BEES?						
QUEEN BEE IS PRESENT / IDENTIFIABLE?						
INFESTATION BY ANTS / ANTS PRESENT?						
INFESTATION BY WAX MOTH / WAX MOTH PRESENT?						
NEGATIVE ODOUR NOTICEABLE?						

Lightning Source UK Ltd.
Milton Keynes UK
UKHW030943141122
412173UK00010B/2017